The Great Meteor Procession

THE GREAT METEOR PROCESSION

Horace A. Smith

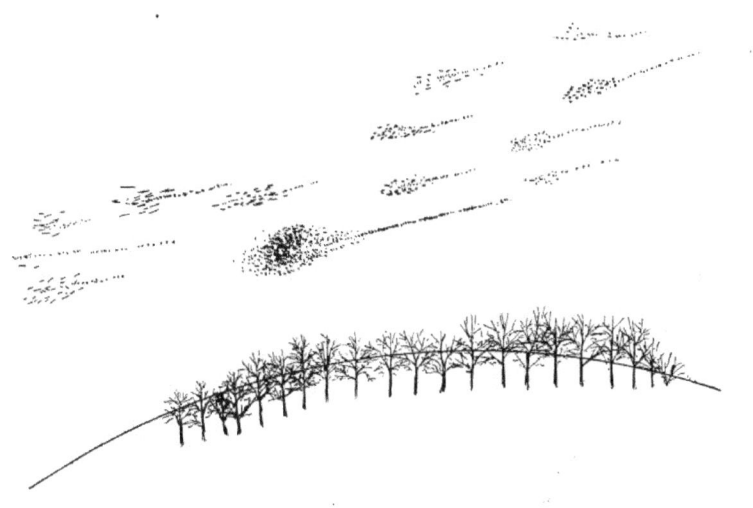

Copyright ©2023 Horace A. Smith
All rights reserved.

Title page: Sketch of the 1913 meteor procession as seen by J. T. Ormiston.

First Edition
ISBN 9798372032767

Table of contents

Preface iii

Chapter 1 Introduction 1

Chapter 2 Chant Investigates 5

Chapter 3 Doubts 37

Chapter 4 Old Newspapers 53

Chapter 5 The Cyrillids 61

Chapter 6 Little Moons? 71

Chapter 7 Earth-Grazers.......................... 81

Chapter 8 Year of Meteors 89

Chapter 9 Meteors or Aliens? 105

Chapter 10 Conclusions
and Loose Ends 117

Glossary 122

Bibliography 124

Index 131

Preface

This book had its origin in a lecture I gave when I taught astronomy at Michigan State University. At that time, I posted a short account of the February 9, 1913 fireball procession on my departmental website. In retirement, I have updated, corrected, and enlarged that earlier document to make this book. I thank Cynthia Ridder, Nancy Silbermann, and Mary Anderson who read a draft and who provided corrections and comments which improved the final book.

The 1913 event holds a particular fascination for me because I live in Michigan. Several reports exist of Michiganders seeing the procession of meteors cross the sky on that chilly February evening. The observer at the Weather Bureau station on what was then the campus of Michigan Agricultural College, but which would become Michigan State University, wrote that at "8:09 PM [Central Standard Time], a number of meteors, probably 15 or 20 in all, crossed the sky about 30° above the northern horizon. The first one seen was a small cluster, such as is seen when a skyrocket bursts; the others followed in quick succession, some singly, others in clusters, like a straggling flock of birds. The rate of motion was apparently about like that of a skyrocket, which they all much resembled. The apparent direction was from about West 10° North. Some of the smaller ones burned out, but the larger ones were lost in the distance in the northeast."

No one now living saw the Great Meteor Procession, but subsequent investigations have given us some understanding of what occurred. With thanks to those who did the investigating – Clarence Chant, William Denning, Alexander Mebane, John O'Keefe, and others – this book tells the evolving story of the Great Meteor Procession.

Figure 1. This photograph depicts the Weather Bureau building erected in East Lansing in 1910. It was presumably from this location that the Great Meteor Procession was seen three years later. From Beal (1915).

Chapter 1
Introduction

On the evening of February 9, 1913, people on the streets of Toronto found their eyes suddenly drawn skyward. A parade of fireballs sailed majestically through the dark sky. As the first blazing group passed, more fireballs appeared, to be followed by still more, as one, then two, and finally three minutes went by. At the end, a roar of thunder rumbled through the chilly night. As all once more became tranquil, amazed eyewitnesses were left with striking impressions. To one watcher the fireballs looked like an aerial fleet engaged in celestial maneuvers. To another, they resembled cars of a distant railroad train journeying through the night. Still another was driven to exclaim "Oh, boys, I'll tell you what it is – an aeroplane race." A perhaps more spiritual witness suggested that the fireballs were souls flying to heaven.

The fireballs were seen by hundreds, if not thousands. Nor was the fireball parade limited to Toronto. Investigators eventually reconstructed the path traversed by the fireballs, a track stretching thousands of miles from western Canada across Ontario and the northeastern United States, and continuing past Bermuda to the south Atlantic Ocean. The fireballs appeared to move almost in single file, leading some to term the display the Canadian Fireball Procession of 1913. However, because the fireballs were also seen far beyond Canada's borders, it is more fitting to use an alternative name, the Great Fireball Procession, or the name I shall most often use, the Great Meteor Procession.

Canadian astronomer Clarence A. Chant called the fireball procession "a meteoric display...quite without parallel." British meteor expert W. F. Denning agreed: He had neither seen nor heard of anything quite like the 1913 display. Charles Fort, who did not think highly of astronomers, nonetheless was sufficiently intrigued by the display to include it in his 1923 book of wonders and mysteries. However, in 1939 meteor expert C. C. Wylie dismissed the whole idea as a mistake. There had been no meteor procession! In the 1950s, chemist Alexander Mebane in turn challenged Wylie's dismissal. He would write many letters and delve into newspaper archives in a quest for new data. Soon thereafter, astronomer John A. O'Keefe also rebutted Wylie, while connecting the fireballs with a fifth century Alexandrine saint. In the postwar world, others would associate the meteor procession not with a saint but with flying saucers.

The Great Meteor Procession itself was of short duration, occupying but a small part of a single night. It was long before the time of camera-carrying cell phones and the fireballs were not photographed. We have today only the written accounts and, more rarely, the drawings left by a few of those who saw and sometimes heard things they found startling and awe-inspiring.[1] More than a century later, the procession remains unparalleled,

---

[1] Many important papers dealing with the meteor procession, listed in the bibliography, are freely available online through the NASA Astrophysics Data System.

although, as we shall see, it is not entirely lacking in kindred celestial phenomena.

In this book I tell the story of the Great Meteor Procession and of the decades-long attempts to fathom its nature and origin. As we proceed through the story of the fireballs of 1913, we look back in time, not just to the year of their flight, but to the early days of meteoric science, more than a century before that. We also leap ahead, leaving 1913 behind, as we go to the dawn of the Space-Age and beyond. Along the way, the reader will encounter hikers, sailors, artists, astronomers, and iconoclasts, and will discover that satellites, UFOs, paintings, religious calendars, and a poem are all part of the story.[2]

---

[2] Note that a short glossary is included, beginning on page 122, for those unfamiliar with terms such as meteoroid and escape speed.

Great Meteor Procession

Chapter 2
Chant Investigates

Clarence Augustus Chant (1865 – 1956) was a pioneer of Canadian astronomy, and it is thanks to him that the event of February 9, 1913 attracted lasting attention. At the time, Chant taught at the University of Toronto where he championed astronomy and had argued, so far unsuccessfully, for the acquisition of an astronomical observatory.[3] To his regret, Chant was inside on that Sunday evening when the fireballs passed, and he did not see them. Radio for the general public was still in the future, but the fireballs made headlines in Monday's dailies. Chant wrote that, although telephone calls had soon alerted him to the passage of the fireballs, it was not until he read the next day's newspaper accounts that he appreciated how extraordinary those fireballs had been.

Intrigued by the newspaper reports, Chant sought further information. He asked local newspapers to publish a call for observations. As Canadians mourned the death of Captain Robert Scott and his party in Antarctica, word of which arrived as the fireballs passed, newspapers complied with Chant's request. His appeal would not be in vain, and letters from witnesses soon began to arrive.

---

[3] Chant's efforts, along with those of others, would eventually lead to the creation of the David Dunlap Observatory, which opened in 1935.

Figure 2. Clarence Augustus Chant (Wikimedia Commons; Public Domain)

FIFTEEN BIG, FIERY METEORS SHOT OVER CITY LAST NIGHT ONE LARGE AS HALLEY'S COMET

People in All Parts of Toronto Viewed Strange Spectacle at Nine o'Clock When Huge Balls of Fire Whizzed Across the City, Followed by Thunder.

LEADING CITIZEN DEAD AT GUELPH

Christian Kloepfer, Ex-M.P.

SCORES OF METEORS ILLUMINATE THE SKY

Hundreds of People Watched a Beautiful Spectacle Lasting Two or Three Minutes

Dublin Suffragettes on Hunger Strike

(Canadian Press Despatch.)
DUBLIN, Feb. 9.—The

Figure 3. Toronto's newspapers for February 10, 1913 reported the passage of the fireballs the night before. Here we see headlines from the front pages of the Toronto World (top) and the Globe (bottom).

The Meteoric Display of Sunday Evening

To the Editor of The World:

The undersigned would consider it a great favor to receive information regarding the meteoric display of Sunday evening. Reports from those living some distance from Toronto would be especially valuable, particularly in regard to the position in the sky in which the meteors were seen—that is, whether apparently overhead or to the east, the west, or any other direction; and if not overhead, how high above the horizon they were.

Kindly give as definite information as possible regarding the following: Time of occurrence, position in the sky, direction of the motion, how many seen in all, how many at once, how long whole phenomenon lasted, if any sound was heard; if so, at what time it was heard and what it was like; if bodies remained intact or broken up; if bodies had tails and how long they were; how long any body was in sight.   C. A. Chant.

University of Toronto, Toronto, Feb. 11.

Figure 4. The Toronto World for February 13 carried Professor Chant's appeal to witnesses of the display.

Figure 5. A grayscale version of eyewitness Gustav Hahn's painting of the 1913 fireball procession. The meteors are depicted streaking over the darkened streets of Toronto, approaching the constellation Orion. The spacing of the meteor groups has been somewhat condensed. A version of Hahn's painting was included in Chant's first paper about the meteors. Wikimedia Commons. Public Domain.

Chant studied the letters which arrived in the post. They came not only from Toronto, but from locations far beyond that city. Letters eventually arrived from the plains of Saskatchewan, the island of Bermuda, and many points in between. However, even those in the same geographic area did not always describe the fireballs in the same way. Some saw more fireballs, some fewer. Some wrote that the display was visible for two minutes, others put it twice as long. Reports of their direction of flight likewise differed. The observers had, of course, been taken by surprise that winter night, and days or weeks may have elapsed before they ransacked their memories to write to Chant.

Making sense of conflicting eyewitness testimony called for careful judgment and sometimes necessitated follow-up inquiries by the professor. When he published his first paper on the subject a few months later, Chant noted that readers would "see that intelligent people can differ widely in describing a phenomenon, and will be able to appreciate the difficulty I have had in discriminating between very discordant observations."

Weighing the varied and sometimes conflicting communications from his informants, Chant strove to merge the reports into a coherent picture. The accounts he received, and the conclusions he drew from them, were published in a long paper in the May-June issue of the *Journal of the Royal Astronomical Society of Canada*.[4] Titled "An Extraordinary Meteoric Display" (Chant 1913a), the paper summarized the extraordinary spectacle:

> GENERAL DESCRIPTION *[As seen in Western Ontario]*
>
> At about 9.05 *[Eastern Standard Time] on the evening in question there suddenly appeared in the northwestern sky a fiery red body which quickly grew larger as it came nearer, and which was then seen to be followed by a long tail. Some observers state that the body was single, some*

---

[4] Chant had founded that journal and was its editor.

that it was composed of two distinct parts and others that there were three parts, all travelling together and each followed by a long tail. The front portion of the body appears to have been somewhat brighter than the rest, but the general color was a fiery red or golden yellow. To some the tail seemed like the glare from the open door of a furnace in which is a fierce fire; to others, it was like the illumination from a search light; to others, like the stream of sparks blown away from a burning chimney by strong wind.

The first suggestion which occurred to many who saw the body was that someone had set off a great sky-rocket. In the streaming of the tail behind, as well as in the color, both of the head and the tail, it resembled a rocket; but, unlike the rocket, the body showed no indication of dropping to the earth. On the contrary it moved forward on a perfectly horizontal path with peculiar, majestic, dignified deliberation; and continuing in its course, without the least apparent sinking towards the earth, it moved on to the south-west where it simply disappeared in the distance.

As we all know, most shooting stars are visible for but a very short time, and the brilliant ones very generally descend rapidly towards the earth, seemingly (as one of my correspondents remarked) 'in a mighty hurry to reach their destination'; but here were bodies moving leisurely along, giving ample time for the fortunate observer to make several wishes if he were so inclined. Some report that just before disappearing this body burst, leaving behind it a trail of stars. Before the astonishment aroused by this first meteor had subsided, other bodies were seen coming from the north-west, emerging from precisely the same place as the first one.

Onward they moved, at the same deliberate pace, in twos or threes or fours, with tails streaming behind, though not so long nor so bright as in the first case. They all traversed the same path and were headed for the same point in the south-eastern sky. Gradually the bodies became smaller, until the last ones were but red sparks, some of which were snuffed out before they reached their destination. Several report that near the middle of the great procession was a fine large star without a tail, and that a similar body brought up the rear.

To most observers the outstanding feature of the phenomenon was the slow, majestic motion of the bodies; and almost equally remarkable was the perfect formation which they retained. Many compared them to a fleet of airships, with lights on either side and forward and aft; but airmen will have to practice many years before they will be able to preserve such perfect order. Others, again, likened them to great battleships, attended by cruisers and destroyers. Should these bodies strike the earth they might prove destroyers indeed! Still others thought they resembled a brilliantly lighted passenger train, travelling in sections and seen from a distance of several miles. The flight of the meteors has also been compared to that of a flock of wild geese, to a number of men or horses in a race, and to a school of fish, startled and darting off in a single direction. These and many other interesting details will be found in the reports of observations printed below.

Just as the bodies were vanishing, or shortly afterwards, there was heard in many places a distinct rumbling sound, like distant thunder or like a carriage passing over rough roads or over a bridge. In some cases three such sounds, following at short intervals, were heard; while a number of people felt a shaking of the earth or of the house.

As to the number of bodies there is great diversity of statement. The usual estimate is from 15 to 20 but some say 60 or 100, while some say there were thousands. Various reasons can be assigned for the discrepancy between these numbers. Those giving the small numbers probably refer only to the chief bodies, and as some people have better eyesight than others, where one would see a single body others would see its different parts. Those who report the large numbers undoubtedly included fragments of the larger bodies and also the many red stars bringing up the rear. The only person that I have heard of who viewed the meteors with any instrumental assistance was Master Cecil Carley, a pupil of the Trenton High School, who used an opera glass. He says: 'There were about ten groups in all and each group, as seen through the *opera glass, consisted of from twenty to forty meteors.'*

The entire time occupied by the display cannot be determined accurately, but is given below as perhaps 3.3 minutes. This is an extraordinarily long time for such a phenomenon, but there is good evidence that it is not an exaggeration. The stretch of country over which the display was seen is also unprecedented. In September, 1868, a fire-ball was traced from over the Black Sea to France, about 1500 miles; and on December 21, 1876, such a body first became visible in Kansas and disappeared near Niagara Falls, thus covering a distance of over 1000 miles; but in the present case persons living 2500 miles (one-tenth of the Earth's circumference) apart saw the same bodies. Moreover the description[s] furnished by observers in Bermuda, in Ontario and in Saskatchewan do not materially differ.

While Chant's description provides an excellent overview of the meteor procession, a better appreciation of the material upon which he based his summary can be gained by looking at just a few of the many reports he received. We turn first to a pair of Canadian hikers, outdoors on that chilly winter evening at Parry Sound, Ontario, some 120 miles (195 km) north of Toronto. One of the pair, Walter L. Haight, recalled what they witnessed.

On the evening in question I happened to be returning from a snowshoe tramp, and was in the act of tightening up the straps on my foot when my companion called out: "Look! Look!" and I immediately threw my head up and caught sight of the large meteor, which appeared to be traveling very slowly – so slowly that the stateliness of its motion excited my liveliest surprise and wonderment...While my gaze was riveted on the large body, and just when it was about passing out of sight, my companion again called out "Look! There and there!" and I looked up and saw the first group of following meteoric bodies...Before I could recover from my astonishment a new group of smaller ones...came sailing along...I likened them at the time, and the resemblance seems yet apt and appropriate to a large battleship moving ahead with attendant squadrons of torpedo-destroyers and torpedo boats.

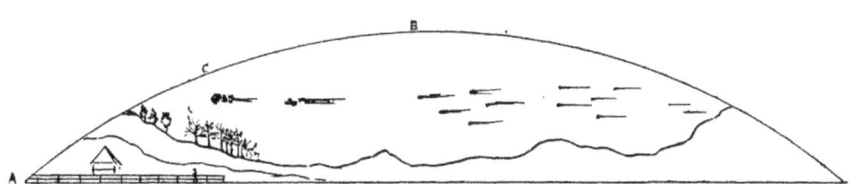

Figure 6. Sketch by Walter L. Haight, observing from Parry Sound, Ontario, north of the city of Toronto.

Great Meteor Procession

Figure 7. Drawing by W. Betts of Toronto. The first meteors were stated to be brilliant golden balls, like fireworks on the verge of exploding. The following meteors were not so brilliant, but were still bright. Betts's reduced the spacing of the meteors to fit them in one drawing.

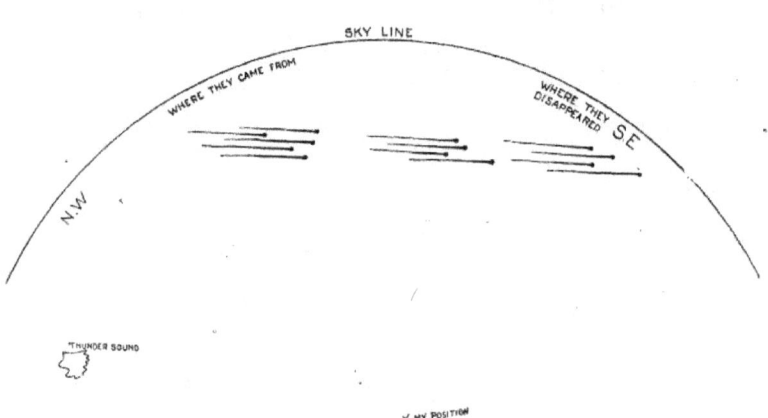

FIG. 4.—The flight of the Meteors, as sketched by Mr. A.W. Banfield at Berlin. The observer was at X and faced north-east. The place where the sounds appeared to come from is indicated in the left foreground.

Figure 8. Sketch drawn by A. W. Banfield, who was at Berlin, some 58 miles from Toronto.

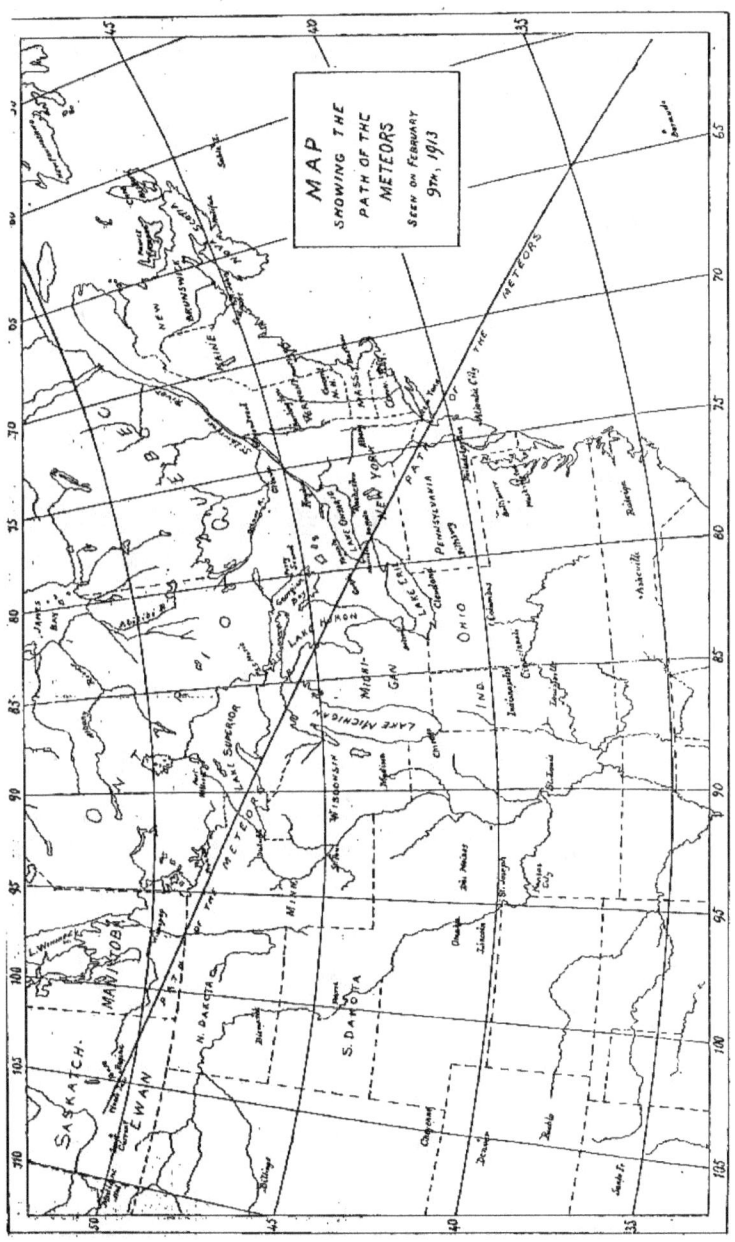

Figure 9. Chant constructed this map showing the track (trace, in Chant's terminology) of the procession.

Great Meteor Procession

James H. Bolton, observing some 23 miles northwest of Toronto, wrote:

When I saw the first meteor at 9.05 it was a little west of north-west from here and travelling nearly toward me. I took it for an aeroplane with both headlights lit, and as it came nearer the sparks falling behind it made it appear still more like one. However, after a minute or a minute and a half I could see it was a meteor, and the tail seemed to cover about half the sky distance when it was passing. It was very low, apparently just above the hills, and by this time I noticed about 12 or 15 more following it. Would guess that it was about 3 minutes from the time I saw the first and largest one until it got out of sight.

The smaller ones were going a little more slowly, and some of them died out just as they passed over, though they were not directly over me, but a little to the southwest. Would think it about 5 minutes from the time of first notice until they were all gone.

About 2 minutes after they disappeared there was a heavy noise like a clap of thunder at a distance. Half a dozen of us were together and all noticed it distinctly; in fact it was too heavy to go unnoticed. There would be 12 or 15 passing at one time, so would think there were about 30 in the whole procession. I have been fortunate enough to see nearly every big meteoric display for the past 50 years, but never saw anything as fine as this.

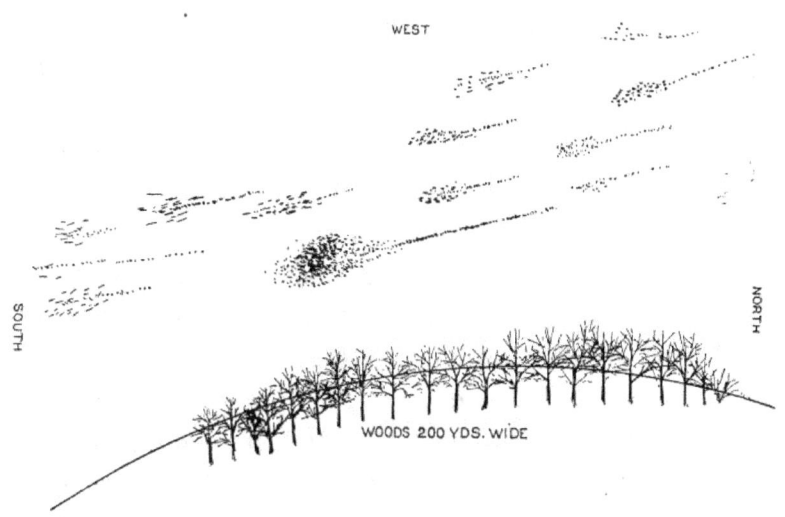

Figure 10. J. T. Ormiston saw the procession from Centreton, Ontario, 74 miles from Toronto, and drew this sketch.

In Bermuda, Colonel W. R. Winter's sighting, while spectacular, lacked the ensuing rumble of thunder heard from locations closer to Toronto:

> *I saw two leading bodies like large arc lights in appearance, slightly violet in color; diameter, as far as the brilliance would permit of judgment, equal to that of the moon[5]; one lower than the other and a little in advance of it. Both were*

---

[5] The waxing crescent moon had set, so that no direct comparison with the moon was possible.

coruscating or breaking into small pieces. As these pieces separated from the parent bodies they developed trails of sparks and gas. There were about 100 of these followers. Each had a nucleus, and perceptibly dwindled away in length as they got behind, from a length of about 13° near the head of the procession to small spots resembling pieces of wind-driven burning brown paper at the end. The longer ones were slightly scimitar-shaped and of a yellowish-red appearance, the nucleus being very bright. As they shortened up the color became more red, until at the last they were quite red, with a bluish flame above—exactly as if they were burning. These latter appeared to drift with the wind. The short ones preceding these were apparently egg-shaped. They were travelling horizontally, with a slight downward tendency as a whole.

Figure 11. W. R. Winter's sketch of the procession as seen from Bermuda.

METEORS AS SEEN IN BERMUDA

Figure 12. George Gosling's view of the lead fireballs, as communicated by W.R. Winter, was included in Chant's second paper published at the end of 1913 (Chant 1913b).

Upon reading Chant's first paper on the fireballs, W. R. Winter wrote to him again. Winter called attention to a difference in the appearance of the meteors when seen from Bermuda compared to Toronto. He wrote that in "Ontario the bodies were arranged in distinct groups, while at Bermuda there were two brilliant ones leading a large number of bodies distributed fairly regularly in a long train behind" (Chant 1913b).

On February 14, 1913, long before Chant's papers appeared, a New York newspaper, *The Evening World*, published an account from the captain of the SS Zafra, newly arrived in New York but northeast of Bermuda on the night of February 9. The headline to the page 6 article related how the Zafra's crew had trembled at a terrifying sight (Figure 13).

METEORS CAP CLIMAX OF FRIGHTFUL INCIDENTS.

Half-clad, Capt. Abbott rushed up to the deck. An unearthly flare hung over the ship, and, sailing across the sky was what looked like a flock of monstrous birds of fire. They were coming towards the Zafra, and they passed over her, shedding their unearthly radiance, at an altitude, Capt. Abbott thought, of about 2,000 feet.

"They'll light on us and burn us all up," yelled Ramon Fernandez, feeling for an amulet that was not there.

But the meteors, while Capt. Abbott was verifying his count that there were forty of them, sailed on to the southwest. The crew stayed on deck, shivering and praying, until the last faint glow of their taillights had flickered away in the distance.

And Capt. Abbott has it all down in the log, from the disappearance of Ramon's amulet, and of Evangeline and Snap, to the coming of the meteors. Lat. 33.45, Long. 63.30.

Figure 13. The story in the New York Evening World for February 14, 1913.

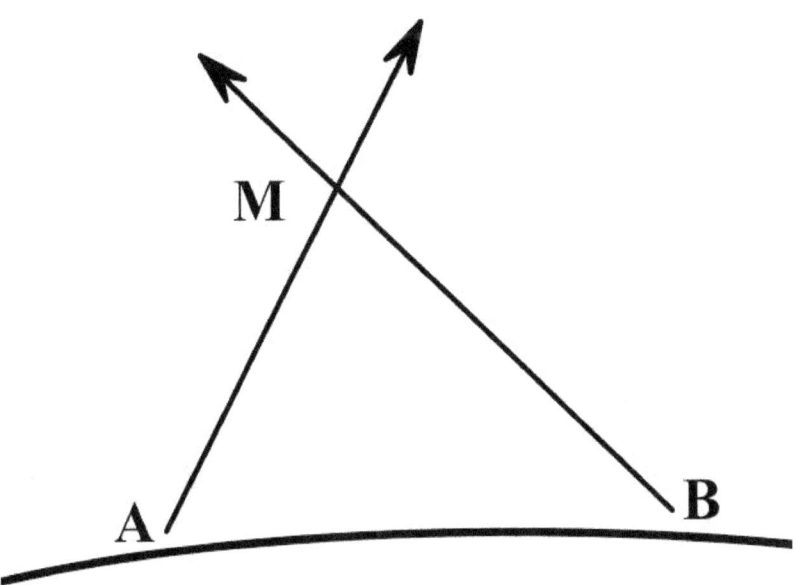

Figure 14. When two observers at known locations A and B record at the same instant the apparent position in the sky of meteor M, the height of the meteor can be calculated. Chant faced a more perplexing problem. Few of the untrained and surprised observers could precisely describe the apparent path taken by the meteors. Nonetheless, Chant did his best with the observations at hand.

Chant attempted to derive quantitative as well descriptive information about the meteors. How high were they and how fast did they move? The question of height could in principle be answered geometrically. If the path of a meteor through the sky is observed from two or more well-defined locations, separated by twenty or thirty miles, then the height of the meteor above the ground can be calculated (Figure 14). Surprised observers of the 1913 fireballs were often unable to report the apparent path of the meteors through the sky as accurately as Chant hoped. Nonetheless, although Chant had difficulties combining contradictory observations to compute the heights of the meteors, he came up with a result. He decided that they were probably lowest, perhaps 26 miles (42 km) above the ground, in the Toronto-Hamilton area, where thunderous sonic booms followed their passage. Taking notice of the greater heights later computed by Denning and Davidson, Chant subsequently raised that height to 34 miles.

Determining the speed of the fireballs posed other difficulties. Observations gathered by Chant traced the path of these fireball procession for an unrivaled 2500 miles (4000 kilometers). As the meteors were seen at places nearly 2500 miles apart, it might be supposed that calculating their speed would be easy. Simply note the times when the fireballs were seen to begin and to end their flight. Subtract the two, taking account of different time zones, to get the duration of the flight. Divide the distance they flew by the elapsed time to get the average speed. All well and good. However, the entire time consumed in travelling 2500 miles was short, only several minutes, while the observed times were themselves uncertain by a few minutes. The time of flight was not known accurately enough to determine the speed of the fireballs in the manner described. Chant was nonetheless able to estimate that the fireballs had a speed of more than 5 and less than 10 miles per second. That is slower than the speeds of most meteors (7-45 miles per second or 11 to 72 km per second) and is in the neighborhood of the 5 mile per second speed of a satellite in a circular orbit not too far above the surface of the Earth. Chant suspected that the meteors did not advance at a constant speed but slowed in going through the upper air.

Putting everything together, Chant arrived at his interpretation of the display: "It would seem that the bodies had been traveling through space, probably in an orbit about the sun, and that on coming near the earth they were promptly captured by it and caused to move about it as a satellite." Here "bodies" refers to the meteoroids which, upon entering the Earth's atmosphere, created the meteors seen from Saskatchewan to Bermuda. The meteors encountered our world at a shallow, grazing, angle.

A typical meteoroid orbits about the Sun; some of them are in orbits which cross the orbit of the Earth. If Earth and meteoroid meet, the latter plunges into the atmosphere at speeds of 7 – 45 miles per second. The minimum speed is that expected when an object, initially at rest with respect to the Earth, falls to our planet from a great distance in a vacuum. The maximum speed comes when Earth and the meteoroid go around the Sun in opposite directions and the pair smash head-on. Satellites in circular orbits with an altitude of a couple hundred miles move at a speed of 5 miles (8 km) per second. Hence, a re-entering satellite, or something resembling one, moves more slowly than most shooting stars.

A meteoroid striking our planet ordinarily starts to glow as a meteor in the tenuous upper atmosphere, 50 - 75 miles (80 – 120 km) above the ground. Altitudes of 60 miles or 100 km are typical for the first appearance of a shooting star. The meteoroid which gives rise to a typical shooting star is only about the size of a grain of sand or perhaps a pebble, although the glow around it is much larger in size. Such a small meteoroid is generally turned entirely to dust and vapor during its brief passage through the atmosphere. Larger meteoroids sometimes survive to hit the ground, becoming meteorites. Most, however, have been greatly slowed by the time of impact.

Meteoroids hit the Earth's atmosphere at different angles, but most will approach in a trajectory heading toward the ground. Occasionally, however, a meteoroid will move on a path almost parallel to the ground below, producing a meteor moving tangentially to the Earth's surface – hence a grazing meteor. A pebble-sized meteoroid on a grazing trajectory can produce a long shooting star trail through the night sky, but it will still usually be destroyed during its passage through the atmosphere. However, interesting possibilities exist for larger grazing objects, such as those which give rise to dazzling fireballs. Some may have only a brief encounter with our world, losing some speed as they move through the air but retaining enough energy to escape Earth and return to the interplanetary space from whence they came (Figure 15).

Suppose, however, that an Earth-grazing meteoroid loses enough energy to reduce its speed below the 8 miles per second escape speed. It is then unable to fly off again into space. Instead, gravity bends its path toward the ground. Sinking into the atmosphere, the meteoroid develops a meteoric glow, but it will continue to be slowed by the air resisting its motion. Its speed will eventually be slowed below the 5 mile per second speed needed to stay in orbit. Then it is either destroyed in the air or falls to the ground, making only a partial circuit of the Earth, as illustrated in Figure 16. In his first paper, Chant proposed this second possibility as an explanation for the fireball procession of 1913.

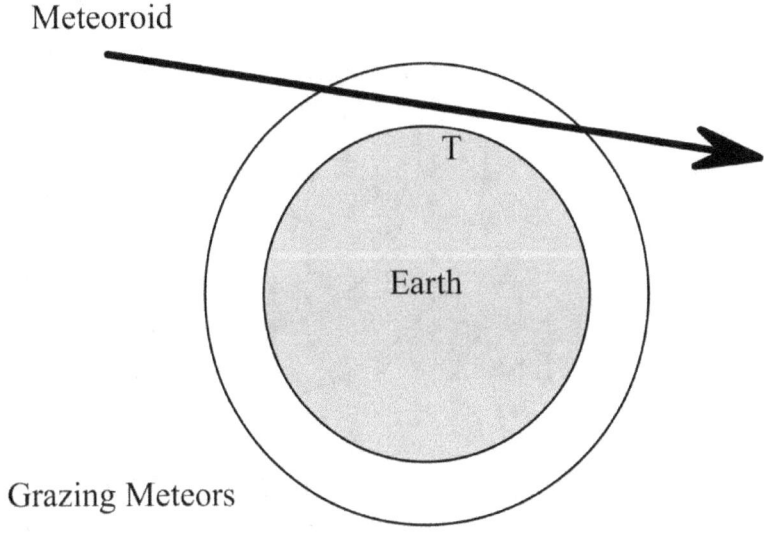

Figure 15. Schematic drawing of an Earth-grazing meteor. The outer circle, not drawn to scale, represents the area of the atmosphere in which the meteor might be visible. The grazing meteoroid enters the atmosphere at a shallow angle, tangential to the surface of the Earth at point T, its lowest altitude. If the meteoroid is moving fast enough, and if the meteoroid does not disintegrate during its passage through the atmosphere, it can escape once again into space – though its motion will likely be slowed and its direction of flight altered to some degree.

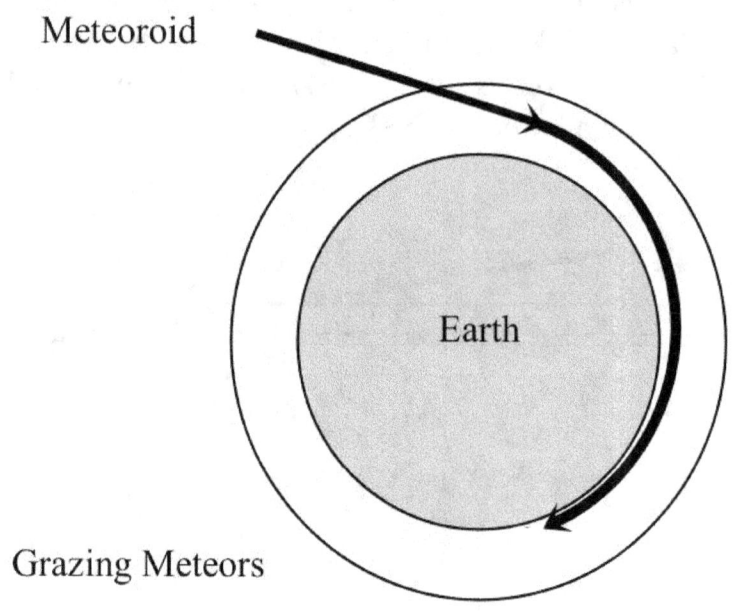

Figure 16. If an Earth-grazing meteoroid is moving at a speed not too much greater than the 5 mile per second orbital speed, it may for a time move as a natural satellite of the Earth. Slowed by air resistance, it will not be able to orbit for long, and the meteor produced can resemble a re-entering artificial satellite. Whether any remnant of the meteoroid survives to reach the ground depends upon its size and chemical composition.

Chant's paper on the meteors attracted immediate attention. Follow-up studies by Chant himself, W. H. S. Monck, M. Davidson, and William Frederick Denning were published in late 1913 and early 1914[6].

---

[6] Full references are included in the bibliography.

Great Meteor Procession

Denning (1848 – 1931) had long been ranked among the most important observers of meteors, even gaining mention in H.G. Wells's *War of the Worlds*[7], where he was deemed "our greatest authority on meteorites". Chant's account of the 1913 meteor procession impressed that seasoned observer. Denning remarked that "I have been in the habit of watching the heavens since 1865 and have never noticed anything similar."

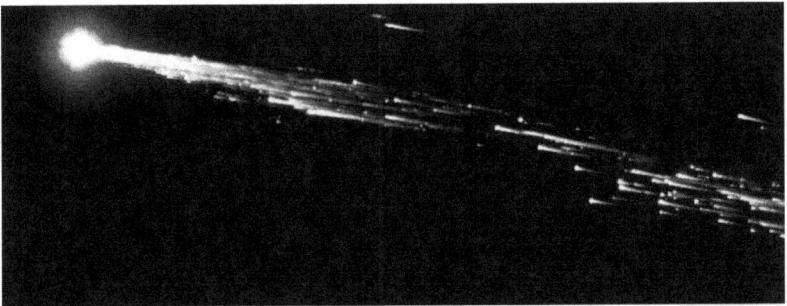

Figure 17. Forty-four years before the launch of Sputnik, none who saw the fireballs in 1913 had seen the re-entry of an artificial satellite. This photograph shows the 2008 blazing re-entry of the European Space Agency's Jules Verne Automated Transfer Vehicle. Courtesy: NASA/ESA/Bill Moede and Jesse Carpenter.

---

[7] Wells's tale of invading Martians was first serialized in 1897 and appeared in book form in 1898. Of the falling star that carried the first invaders, Wells wrote: "Denning, our greatest authority on meteorites, stated that the height of its first appearance was about ninety or one hundred miles. It seemed to him that it fell to Earth about one hundred miles east of him."

Great Meteor Procession

In his 1913 paper, Denning concurred that the Great Meteor Procession had been traced for 2500 miles. From his own analysis of the eyewitness accounts, Denning found that the meteors, when over Toronto, were 38 miles high rather than Chant's 26 miles, and that the meteors were higher still at other portions of their observed path. The speed of the meteors Denning placed at 8 miles per second. He thought that, before encountering the Earth, the meteoroids had been moving around the Sun in the same direction as the Earth, giving the chasing meteoroids a slow relative speed when they met up with our planet.

Denning also noted that the trajectory of the meteoroids could not have been a straight line, of the type depicted in Figure 15. If that had been the case, the meteors would have been 250 miles high at the beginning and end of their observed track, far too high for an atmospheric passage to make them luminous. Instead, the meteoric bodies must have been slowed by the atmosphere and pulled by gravity into something closer to the trajectory shown in Figure 16. Denning believed that at least some of the larger fireball bodies might have survived to make the whole 2500-mile journey, but he thought that smaller pieces disappeared along the way to be replaced by debris torn from the largest meteors.

Since the meteors seemed to be going strong when they disappeared from view in Bermuda, Denning wondered whether they might have flown further. But how could that be established? Might the meteors have been seen by the crews of ships on the Atlantic Ocean? The *Nautical Magazine* published Denning's appeal for observations in their April, 1914, issue, in which he asked

Great Meteor Procession

"whether any one on board ship between the Cape Verde Islands and Bermuda on the night of the 9th of February, 1913, after 10 pm[8], saw anything of a brilliant meteoric phenomenon"?

Almost a year passed, and both Great Britain and Canada were engulfed in World War I, before Denning's appeal bore fruit. A letter arrived from A. Y. Porter, who had been aboard the SS Bellusia[9], who reported that meteors had been seen on the evening in question: "At 10:30 pm I saw the sky lighted up with meteoric fire, starting at north and going by way of east to the southeast…fragments were falling off as it passed along. It looked like red and white liquid fire."

Still later, a second report arrived from W. W. Waddell, first mate of the SS Newlands. The SS Newlands was then off the coast of Brazil, south of the equator, and some 5500 miles from the first sighting of the procession in Saskatchewan, indicating that the fireballs had traversed almost a quarter the circumference of the Earth. Nonetheless, Waddell's account is similar to many from earlier points along the path of the procession:

> *Eight Bells, midnight. – Had just gone on bridge and Fo'castle Head, and I was just about to leave the deck for my bunk when my eye was caught by a bright shooting star in the western sky that traveled away across the heavens at a*

---

[8] The local time would be later for these ships than it would be in time zones further to the west, around Toronto. A satellite in a low orbit would take about 23 minutes to travel a quarter of the way around the Earth.
[9] Olson and Hutcheon (2013) spell the name Bellucia. The SS Bellucia sank in 1917 after being torpedoed by a German U-boat.

height of about twenty degrees above the horizon. As it went it seemed to drop stars for all the world like a rocket when it explodes...I had time to yell out to the second mate, who had relieved me on the bridge, asking if he had seen that, when a whole shower of stars of the same kind came shooting across in the wake of the first one, each of the stars wavering at the same speed and keeping regular distances from one another, all leaving a train of smaller stars in their wake that seemed to be drawn after the parent star (Denning 1916).

Fresh investigations of the meteor procession continued beyond the end of the Great War into the Roaring 20s. Astronomer William H. Pickering[10] (1858 – 1938) discussed the meteor procession in four papers published in *Popular Astronomy* magazine in 1922 and 1923. An inquiry by Pickering led to the discovery by the U.S. Navy Hydrographic Office of three additional probable observations of the meteor procession by vessels at sea. The British steamship Tennyson then between New York and Bermuda, observed, at $9^h 50^m$, "myriads of stars, large and small, passing from North to South, mostly reddish in colour and traveling in a long line," Useful though they were, these new observations did not extend the path of the meteor track beyond the location of the SS Newlands.

---

[10] Pickering's views sometimes departed greatly from the astronomical consensus. He thought, for example, that his telescopic observations provided evidence for life on the moon.

Great Meteor Procession

Although differing with Chant in some details, Pickering concluded that the idea of a great meteor procession was basically correct. He also concluded that the meteors did not have enough energy to escape the Earth. Instead, he thought that they plunged into the south Atlantic Ocean at the end of their long voyage through our atmosphere. Pickering did, however, call attention to the paucity of observations from New York, New Jersey, and Pennsylvania, noting that, while it was partly cloudy in the western portions of New York and Pennsylvania, eastern areas were clearer. Why weren't more observations of the meteors reported from those locations?

Five years after Pickering's last paper on the meteor procession, Harvard lecturer and research associate Willard J. Fisher (1928) introduced two important points. He noted that, if the meteor procession consisted of objects in temporary orbit about the Earth, their path would not trace an exact great circle route over the globe. The rotation of the Earth during the flight of the meteors would cause their path to curve to the west compared to a true great circle. Fisher also pointed out that the equatorial bulge of the Earth would prove an obstacle to the meteors:

> *But a satellite-meteor starting from latitude 50° or 43° and traversing a nearly circular orbit toward the equator would find rising before it the equatorial bulge of the spheroidal Earth, and above this an atmosphere in which it would continually become more deeply entangled...It is not impossible, and the writer is inclined to*

> believe, with Davidson, that the objects seen in Saskatchewan were partly destroyed by atmospheric ablation, partly entangled in the deeper atmosphere over the equatorial bulge and sunk in the Atlantic.; and that the objects which passed the equator were originally higher members of the stream or swarm of meteors, some of which may even have escaped from the equatorial bulge and got away again.

Fisher's paper indicated that, while Chant's interpretation of the procession might be broadly correct, it required adjustment.

Calling attention again to the paucity of reports from parts of the United States over which the procession would have passed, Fisher wondered whether Chant's appeal for observations had been sufficiently widespread to garner attention in that area. As quoted above, Fisher was inclined to agree with Davidson (1913) that those objects which made it past the equator were originally the higher members of a swarm of meteoric bodies, even speculating that some of them may have been able to escape once more into space.

Despite the differences of the various investigators, the 1920s thus ended with what appeared to be a scientific consensus along the lines outlined by Chant in his original paper. The 1933 edition of Robert H. Baker's popular textbook *Astronomy* encapsulated the consensus:

> That meteors sometimes travel in groups was abundantly verified by the remarkable 'meteor procession' of February 9, 1913. According to

Great Meteor Procession

Chant, there were ten or more groups, each containing from thirty to forty meteors, all following the same path. This procession, which occupied more than three minutes in passing above an observer, was visible from western Canada to beyond Bermuda, a distance of 6000 miles. The meteors had long trains; and rumbling sounds, like distant thunder, were heard as they passed.

The next part of the story will show that not everyone accepted that consensus.

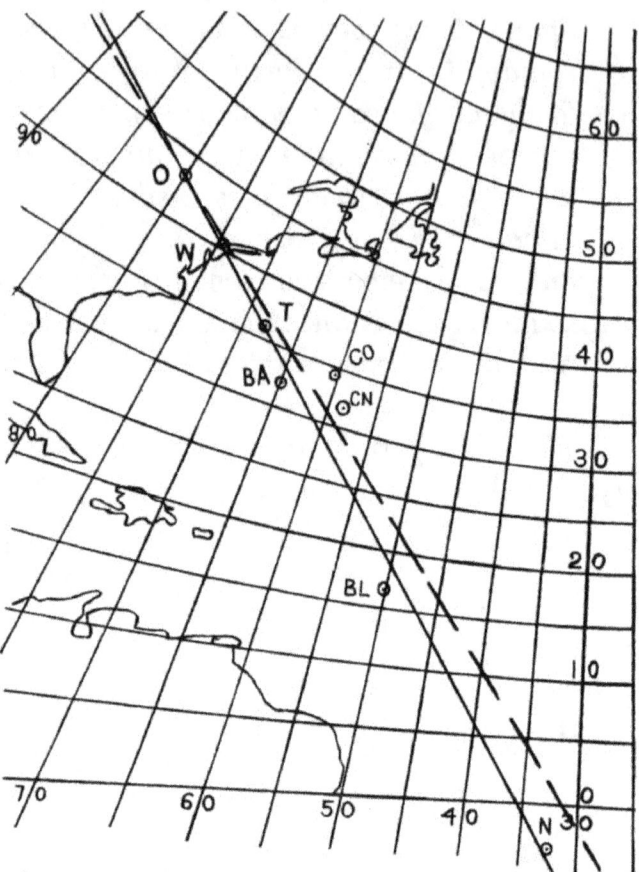

GREAT CIRCLE CHART OF OBSERVERS' STATIONS, 1913 FEBRUARY 9.
O, Ontario station; W, Watchung, N. J.; T, *S.S. Tennyson;* BA, Bermuda; CO, *S.S. Calvo;* CN, *S.S. Custodian;* BL, *S.S. Bellusia;* N, *S.S. Newlands;* solid line, Pickering's great circle; dotted line, Chant's great circle.

Figure 18. Routes of the Great Meteor Procession as drawn by Chant (dotted line) and Pickering (solid line), from Fisher's 1928 paper.

Chapter 3
Doubts

It is an understatement to say that Charles Fort (1874 – 1932) did not accept everything that astronomers happened to write. In his 1919 volume *The Book of the Damned*, he collected accounts of phenomena that he believed could not be explained by science, and which had been unjustly ignored because of that. Astronomers tended to fare particularly badly in Fort's writing.

Figure 19. Charles Fort about 1920. Slightly cleaned from the photo in Wikimedia Commons. Public domain.

In 1923, Fort published another book, *New Lands*, that continued the thrust of *The Book of the Damned*. Fort had already mentioned the 1913 meteors in *The Book of the Damned*, but he took a longer look at them in *New Lands*[11]. In the second book, Fort adopted a decidedly questioning tone:

> *According to data published by Prof. Chant, in the Journal of the Royal Astronomical Society of Canada, 7-148, the most extraordinary procession in our records was seen, in the sky of Canada, upon the night of Feb. 9, 1913. Either groups of meteors, in one straight line, passed over the city of Toronto, or there was a procession of unknown objects, carrying lights. According to Prof. Chant, the spectacle was seen from the Saskatchewan to Bermuda, but if this long route was traversed, data do not so indicate. The supposed route was diagonally across New York State, from Buffalo to a point near New York City, but from New York State are recorded no observations other than might have been upon ordinary meteors, this night. A succession of luminous objects passed over Toronto, night of Feb. 9, 1913, occupying from three to five minutes in passing, according to different estimates. If one will think that they were meteors, at least one will have to think that no such meteors had ever been*

---

[11] The name of Fort's book is an amusing coincidence, given that, in 1923, the meteors were last known to have been seen from the SS Newlands.

seen before. In the Journal, 7-405, W.F. Denning writes that, though he had been watching the heavens since the year 1865, he had never seen anything like this. In most of the observations, the procession is described as a whole — 'like an express train, lighted at night' — 'the lights were at different points, one in front, and a rear light, then a succession of lights in the tail.' Almost all of the observations relate to the sky of Toronto and not far from Toronto. It is questionable that the same spectacle was seen in Bermuda, this night. The supposed long flight from the Saskatchewan to Bermuda might indicate something of a meteoric nature, but the meteor-explanation must take into consideration that these objects were so close to this Earth that sounds from them were heard, and that, without succumbing to gravitation, they followed the curvature of this earth at a relatively low velocity that can not compare with the velocity of ordinary meteors.

If now accepted that again, the next day, objects were seen in the sky of Toronto, but objects unlighted, in the daytime — I suppose that to some minds will come the thought that this is extraordinary, and that almost immediately the whole subject will then be forgotten. Prof. Chant says that, according to the Toronto Daily Star, unknown objects, but dark objects this time, were seen at Toronto, in the afternoon of the next day — 'not seen clearly enough to determine their nature, but they did not seem to be clouds or birds or smoke, and it was suggested that they were *airships cruising over the city.' Toronto Daily Star, February 10 — 'They passed from west to east, in three groups, and then returned west in more scattered formation, about seven or eight in all.'*

Fort has his admirers today. His books are still read, and a magazine named for the iconoclast, *The Fortean Times*, continues to publish reports of oddities and the unexplained. Fort's discussion of the procession of 1913 does not add anything in the way of actual observations to Chant's account. Like Pickering, he raised the question of whether the trail of the meteors from Saskatchewan to Bermuda was continuous, an important issue, which would become the subject of investigations which we will cover in chapter 4. As for the possible daytime observations the following day, the complete account as given by Chant (1913a) is:

In conclusion I shall refer to a curious observation reported in The Toronto Daily Star for Monday, February 10. At about 2pm on that date some of the occupants of a tall building near the lake front saw some strange objects moving out over the lake and passing to the east. They were not seen clearly enough to determine their nature, but they did not seem to be clouds, or birds, or smoke, and it was suggested at the time that, perhaps, they were airships cruising over the city. Afterwards, it was surmised that they may have been of the nature of meteors moving in much the same path as those seen the night before.

Great Meteor Procession

Although the *Daily Star* headlined the report "Fleet of Airships Cruised Over City," it is much too vague a tale on which to hang any weighty conclusion. It implies a single group of observers, as opposed to the many of the night before, and even they do not appear to have been able to see the objects clearly. Any connection with the meteors of the previous evening is weak. One cannot confidently weave this report into the story of the meteor procession itself.

Fort's criticisms of astronomy are often silly. He clearly doesn't understand how the calculations of celestial mechanics are made and tested. Nor has his notion that the Earth is stationary fared well in the Space Age, despite the fancies of latter-day believers in a flat Earth. Some (such as Alexander Mebane, whom we have yet to meet) have stated that Fort interpreted the lights of the meteor procession as a fleet of extraterrestrial spaceships, but he does not actually make so specific a statement in *New Lands*. What he does do is question whether the procession was actually composed of meteors, suggesting instead that people had seen "unknown objects, carrying lights." The nature and capabilities of those objects are not stated.

Charles Fort may have looked askance at astronomers and their beliefs, but his was a voice not likely to be given much credence by scientists. The same could not be said of another critic. Unlike Charles Fort, Charles Clayton Wylie (1886 – 1976) was a credentialed academic. For many years he was a professor at the University of Iowa and his name appears frequently as the author of papers on meteors during the 1930s and 1940s.

Many of Wylie's papers are investigations of brilliant meteors where only chance observations by surprised and untrained observers could be used to trace the path of the objects through the sky. The titles of two of Wylie's papers serve to illustrate the kind of objects with which he was concerned:

The Daylight Meteor of February 19, 1935;
The Detonating Meteor of January 24, 1934.

Wylie's years of sifting observations by the public instilled skepticism in his thinking. Casual eyewitnesses were essential to Wylie's research, but, like Chant, he found that he had to separate wheat from chaff to make sense of their narrations. Wylie summarized the problems that he had encountered in a paper published in 1940. Titled "*Psychological Errors in Meteor Work,*" this paper listed difficulties he frequently ran into when he tried to interpret eyewitness observations of meteoric fireballs. Among the common errors Wylie noted were the erroneous impression that a bright but distant meteor is actually close to the observer and the observer's confusion of the direction in which a meteor is seen with its direction of flight. These common problems perhaps led Wylie to approach reports of the 1913 meteor procession with a jaundiced eye.

In 1937, German astronomer Cuno Hoffmeister wondered whether the 1913 meteors were actually all aligned in a single procession along Chant's track. Could they instead have been members of a world-wide meteor shower? Wylie addressed that question when he took a fresh look at the problem in a paper published in 1939.

Great Meteor Procession

Wylie began by crediting Chant for the initial investigation of the event, but he did not think that Chant had correctly interpreted the observations. Instead, Wylie proposed that observers on that February night had seen the confusing combination of a detonating bolide and a meteor shower. To understand Wylie's paper, it is thus necessary to look at the difference between a meteor procession, as envisioned by Chant, and an ordinary meteor shower.

The reader may be familiar with annual meteor showers, such as the Perseids, which peak each August, or the Geminids, which fly through December skies. The Perseid meteors have that name because their paths through the sky appear to diverge from an area in the constellation Perseus. Likewise, Geminid meteors appear to radiate from the constellation Gemini. The spot in the sky from which the members of a particular meteor shower seem to diverge is termed the radiant. Figure 20, showing Geminid meteors seen in 1914, illustrates this dispersal from a radiant. Shower meteors share a common radiant but they appear to move in different directions through the sky.

Meteoroids responsible for particular meteor showers can often be traced back to specific comets. Comet tails can stretch for millions of miles, but almost all of the mass of a comet resides in a dirty snowball nucleus, a mix of rock and ice perhaps a few kilometers in size. The ices within this dirty snowball gradually sublimate into gases, usually when the comet is closest to the sun, freeing particles that had previously been bound in the nucleus (Figure 21). The meteoroids that produce the Perseid meteors come from Comet Swift-Tuttle. Once every 133 years, Comet Swift-Tuttle comes closer to the Sun than the orbit of the Earth, which last happened in 1992. This is schematically illustrated in Figure 22, although one must remember that it is a two-dimensional representation of what are actually three-dimensional motions. Geminids are somewhat unusual, in that their parent body, Phaethon, looks more like an asteroid than a comet.

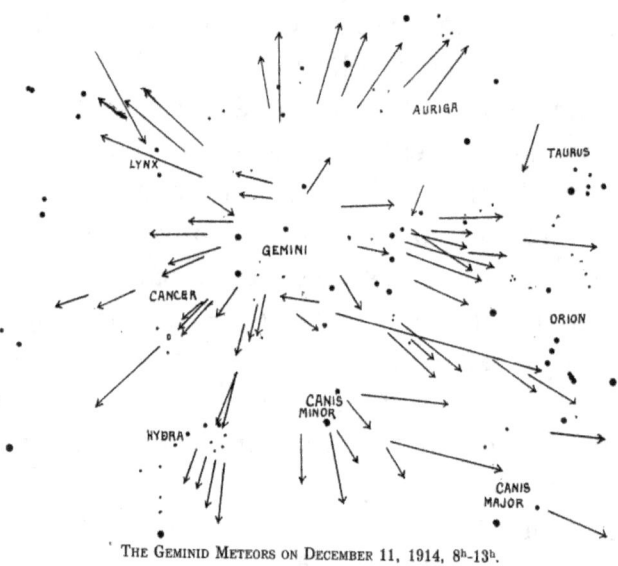

The Geminid Meteors on December 11, 1914, 8ʰ-13ʰ.

Figure 20. Nels Bruseth published this figure of the apparent paths of meteors seen on the night of Dec. 11, 1914. Most, though not all, are members of the Geminid meteor shower and thus their motions trace back to a radiant in the constellation Gemini. (Popular Astronomy, **23***, 100).*

Figure 21. The Rosetta mission obtained this image of the nucleus of Comet 67P/Churyumov-Gerasimenko. A jet of dust and gas can be seen emerging from the nucleus. ESA/Rosetta/NAVCAM, CC BY-SA IGO 3.0.

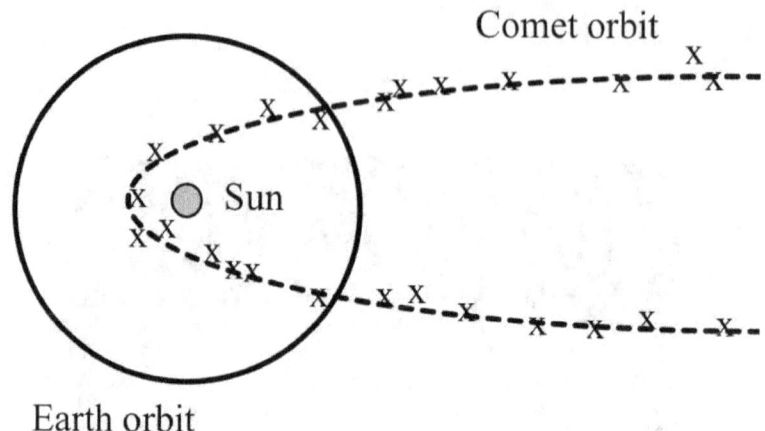

Figure 22. This figure shows the orbit of Earth crossed by the orbit of a comet. Meteoroids, usually small in size, are gradually released from the comet nucleus, spreading out along the comet's orbit, as shown by the x's. When Earth reaches the vicinity of the comet's orbit, it can run into these meteoroids, causing a meteor shower even though the comet itself can be far distant.

When they are freed from the comet nucleus[12], the meteoroid particles continue to move in orbits similar to the orbit of the comet itself, spreading out around and along the comet's orbit. Thus, while Comet Swift-Tuttle has an orbital period of 133 years, we run into material from the comet every August, not once every 133 years.

---

[12] Not all meteors are associated with comets or meteor showers. Most of the fireballs known to have dropped meteorites to the surface of the Earth were singular events. The orbits inferred from the flights of some bright fireballs indicate that they came in on orbits extending out to the asteroid belt between Mars and Jupiter.

Great Meteor Procession

The apparent divergence of shower meteors from a radiant is a trick of perspective. The meteors which seem to go in all directions are actually moving on nearly parallel courses through space. This is similar to the way in which the two parallel rails of a train track appear to diverge from a point in the distance.

Although the Perseids in August and the Geminids in December are usually the richest of the annual meteor showers, an impatient watcher might be tempted to call them meteor drizzles. A stargazer with clear and dark skies might see perhaps fifty or sixty meteors each hour, and even that requires that the shower radiant be well above the horizon. Meteors can trickle from the radiant for hours or days.

On rare occasions meteor showers are not mere drizzles but meteor storms. For example, viewers of the Leonid meteor shower usually see a meager 5 or 10 meteors per hour at its November peak, coming from a radiant in the constellation Leo. Every once in a while, Earth moves through an especially dense stream of Leonid particles and, for an hour or two the number of meteors climbs to thousands an hour. Great Leonid meteor storms were visible from parts of the United States in 1833 and 1966. A lesser but still strong display was visible more recently in 2001. As with ordinary meteor showers, the meteors in a meteor storm appear to diverge from a radiant point in the sky.

The reader will by now have noticed that the radiating pattern of meteor flight in a meteor shower or storm is very different from Chant's meteor procession. On the evening of February 9, 1913, successive groups of meteors followed one another in essentially single-file fashion, or so Chant proposed. Moreover, the procession took only a few minutes to pass.

Figure 23. The great Leonid meteor storm of November 13, 1833 as depicted in Bible Readings for the Home Circle. Library of Congress.

Wylie concluded that Chant's meteor procession did not happen. In his view, there had been two distinct meteoric phenomena that combined to mislead Chant and others. There was a low detonating and fragmenting fireball to the west of Toronto, which gave rise to the thunderous noises heard there. That detonating meteor was, however, seen only in the Toronto area. There was no meteor procession stretching from Saskatchewan to the tropical Atlantic. What, then was seen from the other locations? There was, in Wylie's view, simultaneously an unexpected shower of meteors diverging from a radiant in the north-northwestern part of the sky. It was possible that the detonating meteor over Ontario may have belonged to this meteor shower, but the shower meteors did not follow a single long procession through the sky as advocated by Chant.

Wylie made several points in his debunking of the meteor procession. He noted the lack of observations reporting fireballs rising above and then setting below the horizon, as might have been expected from a fireball procession that traveled across the entire sky. He also pointed out that only low meteors, perhaps 25 or 30 miles above the ground, were known to produce audible explosive sounds. Such a low meteor, he thought, could only travel 8 or 10 miles against air resistance, and thus the detonating meteor observed from Toronto could not have traveled far beyond that city. His map of the motion of this detonating fireball suggested that it did not go beyond the western edge of Lake Erie (Figure 24). Wylie even put forward a probable orbit about the Sun for this detonating meteor before its final encounter with our planet.

Wylie would return to the 1913 meteors in a second paper published fourteen years later in *Science* (Wylie 1953a) titled "Those Flying Saucers". Besides debunking flying saucer reports, Wylie reiterated his view that Chant's interpretation of the 1913 procession was

Great Meteor Procession

erroneous. He continued to argue that there was no procession of meteors tracing a route covering several thousand miles:

> To show the exaggeration possible in an apparently well-authenticated story, consider the story of the fireballs which appeared over the Regina area of Saskatchewan, Canada, on the evening of February 9, 1913, and moved southeastward across Canada and the United States passing nearly over Winnipeg, Toronto, and other important cities including New York City, thrilling and startling thousands of persons in the United States and Canada. This story has been featured 'in several recent magazine articles, for example, giving the number of fireballs passing along that path as 200 to 400 (Coronet, XXXIII, No. 5, 131-132 [1953])[13] and stating that if the fireballs had come to Earth earlier, instead of plunging into the Atlantic Ocean, they would have spread fire and flame over the densely populated area between New York City and Philadelphia...
>
> Obviously what really happened was a shower of shooting stars which was exceptionally good in the Toronto area, but attracted relatively little attention elsewhere. The only report from the densely populated New York City area was from a lady who watched the sky for a while. and counted seven shooting stars. The popular story is impossible of course; and it is evident that an

---

[13] The article was dramatically titled *The Night New York Escaped Disaster.*

excellent but unpredicted shower of shooting stars has been "blown up" into a marvelous procession of fireballs.

Wylie's 1939 paper was influential. J.H. Pruett, writing thirteen years later in the March 1952 issue of *Sky and Telescope* magazine accepted Wylie's conclusions. In a letter to *Science*, answering a criticism of Wylie's 1953 paper by Alexander D. Mebane (who will figure importantly in the next stage of the story), Wylie noted that his suggestion of a new article on the meteor procession was turned down by the editor of *Sky and Telescope* on the grounds that another article was not needed yet but that "Perhaps in a couple more years it would be interesting to remind people of the situation once again" (Wylie 1953b). Wylie's interpretation seems to have gained the upper hand by the early 1950s, but soon two new investigations would challenge Wylie's findings.

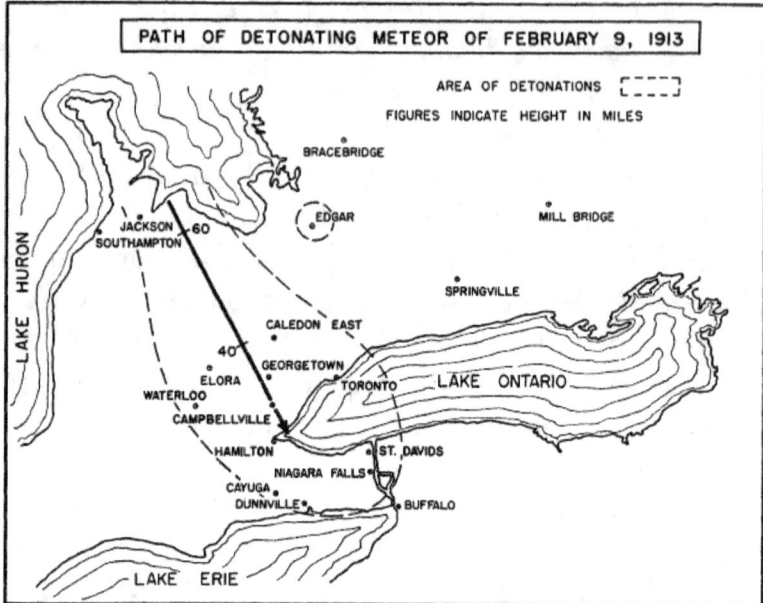

Figure 24. Wylie's (1939) diagram showing the path of the detonating meteor seen in the Toronto-Hamilton area.

Chapter 4
Old Newspapers

Not everyone was persuaded by Wylie's debunking. Lincoln LaPaz did not think that Wylie's rejection of Chant's procession was necessarily right. LaPaz (1897 – 1985) was, like Wylie, a university professor. He joined the University of New Mexico after the Second World War and was the founder and first director of its Institute of Meteoritics. By the early-1950s, LaPaz and Wylie had already crossed swords over scientific issues relating to meteorites, and he was not satisfied to let Wylie have the last word on the meteor procession.[14]

A common objection to Chant's procession was the absence of observations from important regions along its supposed flight path. The skies over the northeastern United States were at least partially clear, so why were so few observations known from that region? The paucity of reports from the densely populated northeastern United States bolstered Wylie's conclusion that the fireballs seen in the Toronto area were a local phenomenon. If additional observations could be discovered along the route that the meteors would have taken as they traversed New York, Pennsylvania, and New Jersey, then the argument for a continuous meteor procession would be

---

[14] Among other disagreements, Wylie and LaPaz disputed whether certain meteorites came from outside the solar system. LaPaz thought they did.

greatly strengthened. But how could such observations be found four decades after the event occurred?

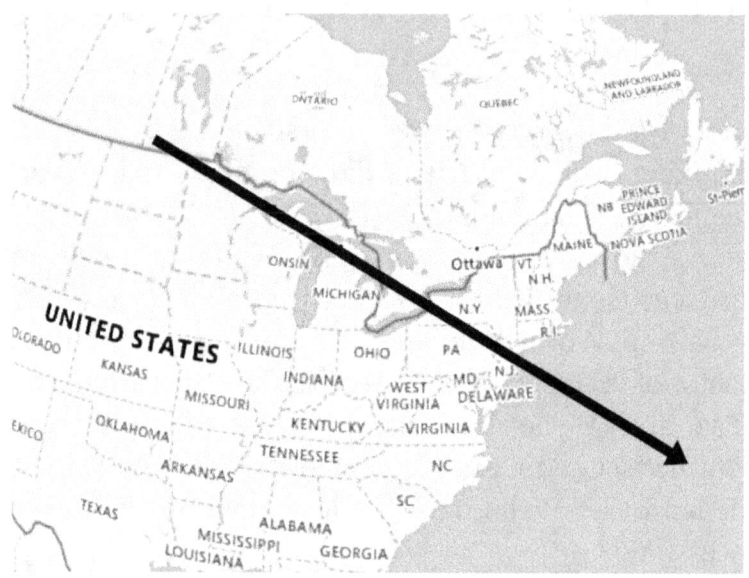

Figure 25. *The arrow on this map shows the general track of the 1913 meteor procession as described by Chant. Mebane would seek undiscovered observations which might fill in gaps along the track.*

One could issue a new appeal for observations, but, while Chant's appeal worked in 1913, could eyewitnesses be expected to recall reliable details after forty years? Even if memories had faded, perhaps there were written records from the time of the event awaiting discovery. Accounts of the meteors, unknown to previous researchers, might sit buried in old newspapers. Digging out such reports would require much labor, especially in that pre-internet age. Somebody would have to take up the challenge of this work. Enter Alexander (Lex) D. Mebane (1923-2004) of New York City.

Mebane was not an astronomer. He graduated from Harvard cum laude in 1944 and obtained a master's degree in organic chemistry from New York University. I do not know how Mebane became interested in the meteors of 1913 or how he became acquainted with LaPaz. We will see in chapter 9 that Mebane had more than a casual interest in the post-war UFO phenomenon, being one of the founders of the Civilian Saucer Intelligence group of New York in 1954.

Mebane set to work, communicating with 615 newspapers located near Chant's great circle route in states from North Dakota to New Jersey, asking local editors to search their files for relevant information. LaPaz allowed Mebane to use the name of the Institute of Meteoritics in his mailings, giving an academic imprimatur to Mebane's efforts. 274 editors answered Mebane's appeal, but their replies provided positive results in only 27 instances.

Mebane did not stop there. He wrote to weather bureau offices in the relevant states and approached local historians in New York state. In New York and Pennsylvania, appeals were published in several newspapers asking for recollections of the, by then, four-decade-old event. All that took a lot of work, and Mebane noted that the "results of this extensive canvass have been somewhat meager in view of the effort expended..." Nevertheless, while Mebane's results might have seemed meager to him after all his hard work, those few positive responses were important. His findings were presented at the September 1955 meeting of the Meteoritical Society and published in their journal *Meteoritics* in the following year (Mebane 1956).

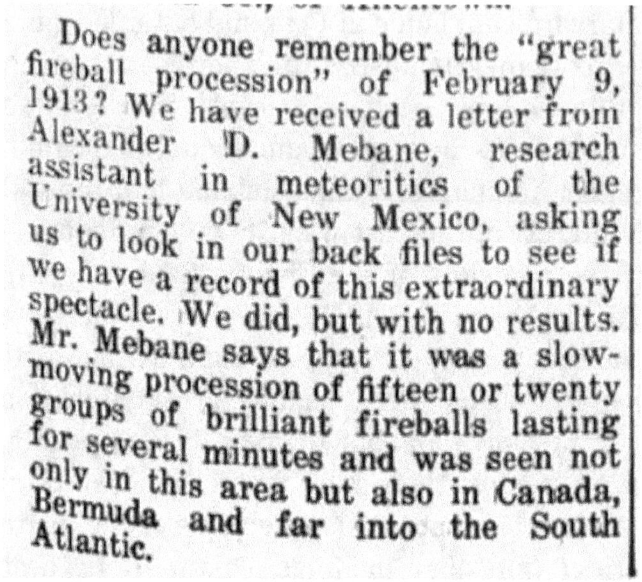

> Does anyone remember the "great fireball procession" of February 9, 1913? We have received a letter from Alexander D. Mebane, research assistant in meteoritics of the University of New Mexico, asking us to look in our back files to see if we have a record of this extraordinary spectacle. We did, but with no results. Mr. Mebane says that it was a slow-moving procession of fifteen or twenty groups of brilliant fireballs lasting for several minutes and was seen not only in this area but also in Canada, Bermuda and far into the South Atlantic.

Figure 26. The Allentown, New Jersey, Messenger-Press published this note in their April 1, 1954 edition – hopefully without thoughts of April Fool's Day.

Mebane's inquiries turned up previously unknown observations of the February 9, 1913 meteors from Minnesota, Michigan, New York, Pennsylvania, and New Jersey, although in Pennsylvania and the New York City area the observations were still scarce. Nonetheless, the additional observations did much to confirm that the meteor procession was indeed continuous across the northeastern United States, answering the criticisms of Wylie and Fort.

The newly uncovered reports were sometimes brief, but a number of them contained details reminiscent

Great Meteor Procession

of the accounts which Chant had collected in 1913. The Elmira (New York) *Advertiser* for February 11, 1913, provided this account:

> "Mr. and Mrs. J. R. Casterline witnessed a strange phenomenon about 9 o'clock Monday [sic] night. They heard a sound like the rumble of thunder, followed by a flash like lightning. They went to the window and saw what appeared to be a shower of falling stars passing across the sky from northwest to south, high above the horizon. The queer part, Mr. Casterline said, was that the stars passed very slowly. They left sparks and trails of fire behind them, and there were sounds like revolver shots. The exhibition continued for 5 minutes or more, until 10 or 12 stars had passed slowly across the sky and vanished one after another in the south, 2 or 3 being in sight at once, part of the time."

Accounts from the New York City area were particularly sparse, but the Plainfield (New Jersey) *Courier-News* noted:

> "As the congregation was leaving the musical service last evening at Watchung, the members were astonished to see 7 stars shooting across the sky from northeast to southwest. They followed each other in rapid succession, and 5 of them passed out of the line of vision, but the remaining 2 seemed to explode like skyrockets and burst into

> *a thousand tiny stars. The 7 lights resembled balls of fire rather than stars."*

The direction of the meteors in that report was not the expected northwest to southeast, but otherwise the sighting reminds us of the procession as seen elsewhere.

The new reports showed that the thunderous sounds associated with the procession were heard in southern Ontario, northwestern New York state, and portions of Pennsylvania. Near Nunda, New York, "*the atmospheric muttering lured more than one man to the barn to see what the horses were about.*" Sounds were not heard over the entire path of the procession, but they extended well beyond the vicinity of Toronto. That would not be expected if they emanated from a single detonating fireball over Ontario, as proposed by Wylie.

Mebane concluded that Wylie's hypothesis would not explain the observations. Whatever happened on the night of February 9, 1913 was much more like Chant's meteor procession than Wylie's meteor shower supplemented by a detonating fireball. Mebane did, however, conclude that the same fireballs were not seen from all locations along the track of the procession:

> *The well-developed procession seen in Hibbing, Minnesota, is surely not identical with the body seen at Houghton, Michigan, which did not break up until near the zenith; and this in turn, tho probably identical with the swarm of sparks seen at Escanaba, can hardly be identical with the 2 large fireballs observed across lower Michigan. The single-file procession across Alpena must be*

distinguished from the less regular procession (the 2 large fireballs?) that passed to the south of Alpena; and since both of these evidently crossed the lake to Ontario, it must be suspected that Chant erred in fitting all of his observations to a single line. On the other hand, it seems possible that all of the New York observations from Buffalo to Elmira refer to a single group of fireballs, whereas the New Jersey observation at Watchung must relate to a different group, since no sound was heard.

The question of sound in itself shows that the data cannot be accommodated in the original framework of horizontally-moving satellitic meteors; there must have been some falling downward, even if the angle was so shallow as to be not easily noticeable by the observers. The westernmost data, at Bemidji and Bovey, mention the thundering sounds of air-racked meteorites; it is not certain whether this phenomenon was noticed at Hibbing, but, if all the observations refer to the same display, we can assume that it was. Yet all across Michigan nothing of the sort was heard, and explosive sounds began to be noticed for the first time only 35 miles west of Hamilton, Ontario; thereafter, they were conspicuous at least to Elmira, and they were probably the same sounds heard 50 miles farther on in Pennsylvania.

Mebane's *Meteoritics* paper was preceded by a short paper by Lincoln LaPaz (1956), scathingly critical of

Wylie's paper. The first sentence in the abstract to LaPaz's paper states "This paper calls attention to the unscientific procedures by which C. C. Wylie has sought to controvert C. A. Chant's satellitory interpretation of the great Canadian fireball procession of February 9, 1913." LaPaz accused Wylie of cherry-picking and fudging the observations so that they provided better support for his alternative to Chant's interpretation. Scientific journals are often reluctant to allow authors to express disagreement so strongly. The circumstance that LaPaz's daughter was associate editor of the journal and that LaPaz was, at the time, closely associated with editor Frederick Leonard, may have eased the paper's route to publication.

LaPaz's disagreements with Wylie were by no means limited to Wylie's interpretation of the 1913 event, and one suspects that a personal animosity may peek through in LaPaz's words[15]. Be that as it may, whatever confusions remained, the Great Meteor Procession in some form was again possible, even probable. LaPaz recommended Mebane's paper as suitable for "further reading" in the popular book on meteorites, *Space Nomads* which he coauthored with his daughter, Jean LaPaz, in 1961. The revived Great Meteor Procession would soon gain additional support.

---

[15] In the late 1940s and 1950s, the Meteoritical Society was apparently riven by antagonisms (Marvin 1993).

Chapter 5
The Cyrillids

What do the meteors of 1913 and a 5th century saint have in common? The association originates with John A. O'Keefe (1916 - 2000), a planetary scientist who joined fledgling NASA in 1958. In a 1959 paper published in the *Journal of the Royal Astronomical Society of Canada*, O'Keefe took up the question of the nature of the 1913 meteors, looking at the concerns that had led Wylie to oppose Chant's natural satellite interpretation. O'Keefe, like Mebane and LaPaz, would conclude that Wylie's arguments were weak or wrong and that Chant's view was closer to the truth.

In his 1913 paper, Wylie identified the location in the sky from which he thought the meteors radiated. However, O'Keefe thought that there was an important problem with Wylie's proposed radiant. It would have been below the horizon as seen from the location of the most southerly shipboard observations. Usually, members of a meteor shower are not seen when the shower radiant is below the horizon, although a few could possibly be visible with the radiant only slightly out of view.

It was unlikely, under those circumstances, that meteors coming from Wylie's radiant would have been so clearly visible to the Bellusia or the Newlands. Since crews aboard those ships did see the meteors, O'Keefe rejected Wylie's radiant. His attempts to adjust Wylie's orbit for the bodies in the meteor shower were unsuccessful and led to other contradictions. Instead, he determined that a nearly-circular, satellite-like orbit provided a satisfactory fit to the observations over the entire length of the sightings.

O'Keefe also noted that each observer saw the meteors form a narrow train as they crossed the sky. if Wylie's ordinary meteor shower explanation were correct, that would require that "the trains seen at different points be actually different trains, so that the whole shower would have to consist of a set of parallel filaments. This raises the question why the observers in one place did not see the filaments observed in other places, as well as their own filament."

Wylie had objected that no one had reported seeing the meteors of the procession rise above or set below the horizon. When Wylie wrote in 1939, no artificial satellites had been launched. However, Sputnik launched in October, 1957, and was soon followed by other satellites. By 1959, O'Keefe could note that observations of rising or setting were rarely reported even for bright artificial satellites. O'Keefe did suggest, as had Mebane and others, that the meteors seen at one location were not necessarily the same as those seen from other spots. There might be a meteoroid swarm, with only the lower members of the similarly moving group being incandescent at any given time.

O'Keefe (1961) suggested that there had originally been a single meteoroid, which had been captured into orbit about the Earth, and that pieces had begun to separate from it at a low point in its orbit. Perhaps on the following orbit, the fragments formed the meteor procession. We will return to this idea in the next chapter.

Great Meteor Procession

If one accepts that the meteor procession consisted of meteoroids temporarily inserted into something resembling a low Earth orbit, could those objects have completed an entire circuit around the Earth? After all, the procession seemed to be going strong when it disappeared from view from the Newlands and Bellusia.

A satellite in low Earth orbit takes about 90 minutes to complete a single orbit. During that time, Earth will spin about 23 degrees. Had the procession returned for a second pass, an hour and a half later, it would have crossed over areas of Canada and the United States about 23 degrees west of the original track. It was in those areas that O'Keefe sought evidence of the procession.

O'Keefe (1961) followed Mebane's lead and began to dig into old newspaper archives. Some newspapers he searched on his own. He hired individuals in Nebraska, Iowa, and South Dakota to investigate local newspapers. Old Weather Bureau reports were also scrutinized. However, the extensive searches uncovered no sightings of the meteor procession (Figure 27). Thus, there is no evidence that any bodies completed a full orbit of the Earth at heights low enough to produce meteors.

O'Keefe also dug into newspapers from locations closer to the Chant track, discovering a few accounts missed by editors to whom Mebane had written. O'Keefe (1968) later examined newspapers in the Canadian provinces of Alberta, Saskatchewan, and Manitoba, discovering only a single new account of the procession in the *Didsbury Pioneer* (Alberta) for February 12, 1913. The Didsbury observation pushed the meteor track some 500 kilometers to the west of the previously known most western point of observation. O'Keefe noted that even further west, in British Columbia, twilight might have dimmed any meteors. However, were they as bright as they were near Toronto, one wonders whether a few might have been visible even before full night fell.

Figure 27. This map from O'Keefe (1961 and 1963) shows the result of his and Mebane's searches for newspaper reports of the 1913 meteors. Dots represent negative results while triangles indicate positive observations of the meteors. Courtesy AAAS.

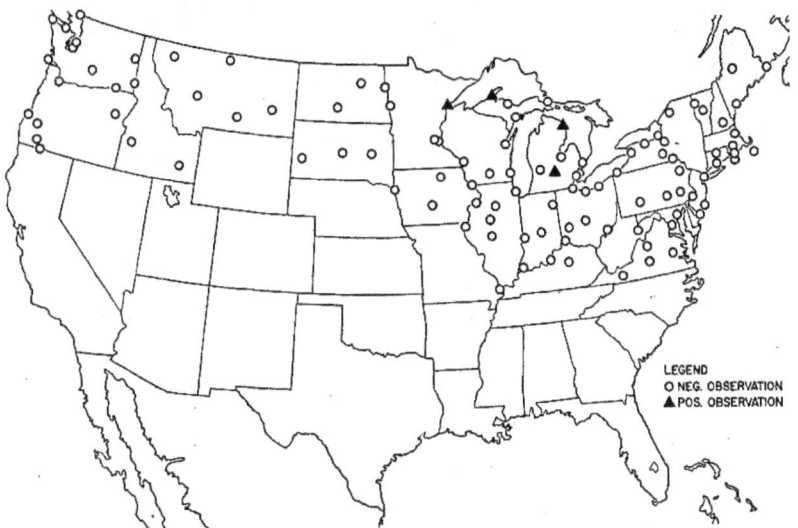

Figure 28. The only positive reports (filled triangles) from the U.S. Weather Bureau were close to the Chant track (O'Keefe 1961). Courtesy of the American Association for the Advancement of Science.

O'Keefe (1960) ventured to give a name to the meteor procession. He noted that the August Perseids were sometimes called the Tears of Saint Lawrence, since August 10 is that saint's feast day. February 9 was the feast day of Saint Cyril of Alexandria. The meteors of the meteor procession therefore became the Cyrillids.

O'Keefe (1963) also wondered whether the Cyrillids might have a connection with a type of meteorite known as tektites. Tektites are glassy rocks found strewn across certain areas of the Earth. O'Keefe considered whether tektites might have been produced by objects like those responsible for the meteor procession, although there were no meteorites found after the passage of the

1913 procession. O'Keefe (1991) later suggested that tektites might have a lunar origin, and that the Cyrillid fragments might be remnants of a ring of material once tossed from the moon. However, most researchers have not been persuaded of O'Keefe's lunar hypothesis for the origin of tektites. Some do, however, believe that tektites have a meteoric connection, but with the Earth rather than the moon. When very large meteoric objects strike the Earth at high speed, molten material can be tossed skyward in the consequent crater-producing explosion. Tektites are, under this hypothesis, ejecta resulting from such impacts which have fallen back to the ground.

Figure 29. Saint Cyril of Alexandria by Jacques Callot (1592–1635). Metropolitan Museum of Art. Joseph Pulitzer Bequest, 1917. Public domain. January 28 is an alternative date for honoring Saint Cyril.

Great Meteor Procession

Alas, O'Keefe's naming of the procession meteors after Saint Cyril was spoiled when a 1969 revision by the Roman Catholic Church bumped St. Cyril's feast day from February 9 to June 27.

As the 21st century arrived, after the labors of Chant, Denning, Pickering, Mebane and O'Keefe, one could be forgiven for thinking that all significant observations of the 1913 meteors had been discovered. Donald W. Olson and Steve Hutcheon thought otherwise. A Texas State University professor, Olson had acquired a reputation as a celestial sleuth well before the centenary of the 1913 display. He, together with students and colleagues, used astronomy to attack mysteries associated with literature, history, and art (Olson 2013, 2018, 2022). As the hundredth anniversary of the meteor procession approached, Olson and Hutcheon reopened the search for undiscovered observations, focusing on ships at sea.

Assisted by archivists in Britain and Germany, Olson and Hutcheon (2013) checked collections of ship's logs for February, 1913. Their findings were presented in an article in the February, 2013, issue of *Sky and Telescope* magazine, in which they reported the discovery of seven previously unnoticed shipboard sightings of the meteors, six uncovered by German archivists and one by a British archivist.

The log of the sailing ship Berthold Vinnen, located off the coast of Brazil, reported that a large number of shooting stars appeared north of Leo, slowly crossing the sky until they disappeared near the stars Alpha Crucis and Beta Centauri. That would indicate that the meteors passed almost over the Berthold Vinnen, moving in a generally south or southeasterly direction. Observations from another sailing ship, the four-masted barque J. C. Vinnen, at 24°29' west longitude and latitude 14°41' south, extended the track of the procession still further. Chant's original 2500 miles track had now grown to more than 7000 miles (11,000 km). In going from Didsbury, Canada, to the J. C. Vinnen, the observed path of the fireballs spanned three-tenths of the Earth's circumference. When last seen from the J.C. Vinnen, the flight of the meteors continued without any sign of an end.

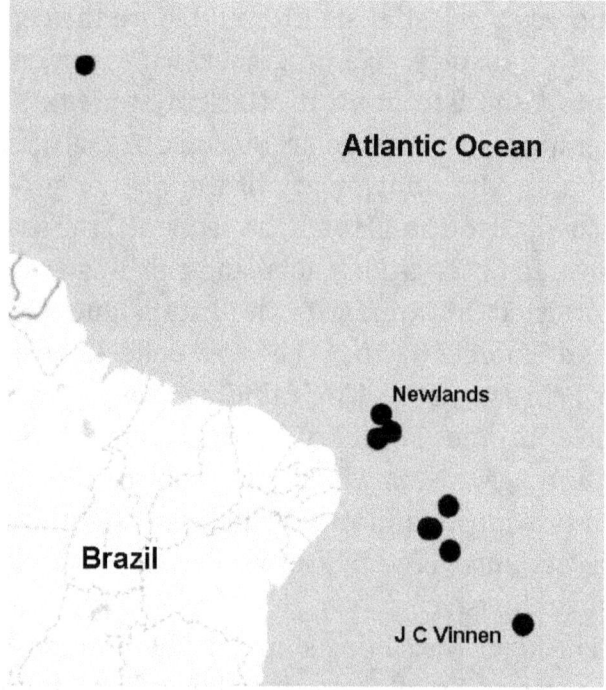

Figure 30. The ships from which the meteors were seen delineate the path taken by the Great Meteor Procession.

It is of interest to compare the first and last eyewitness reports of the procession. The *Didsbury*

Great Meteor Procession

Pioneer in the Canadian province of Alberta provided the first report of the meteors: "Mrs. L. A. Loveland and others reported witnessing a beautiful sight in the heavens on Sunday night about 7:15 p.m. It consisted of from ten to fifteen shooting stars which sent myriads of diamond-like smaller stars from west to east." Our final eyewitness report comes from the J. C. Vinnen, then located in the Atlantic Ocean, south of the equator. The ship's log states that, at 12:40 am local mean time, more than a hundred meteors moved through the sky from Orion to the southern cross (Crux), which would imply a northwest to southeast path. And on they went.

Olson and Hutcheon sought observations still further along the track of the procession. None have been found, and how far the procession continued remains open. A satellite in a low orbit would have taken about 27 minutes to go from Didsbury to the J. C. Vinnen. The times at both locations are only approximate, but, accounting for differences of time zone, we have a travel time of roughly 25 minutes, consistent with the expected transit time of a satellite.

Clarence Chant, whose 1913 paper did so much to bring the Great Meteor Procession to scientific notice, died on November 18, 1956, the year in which Mebane's paper was published and before O'Keefe's paper appeared. I do not know whether Chant learned of Mebane's (and LaPaz's) defense of his explanation for the procession before his death. Unfortunately, the 1913 meteors are not mentioned in his unpublished autobiography, which does not include any events in 1913, jumping from a 1910 solar conference to a 1918 solar eclipse.

Great Meteor Procession

In the next chapter, we look more deeply into natural Earth satellites.

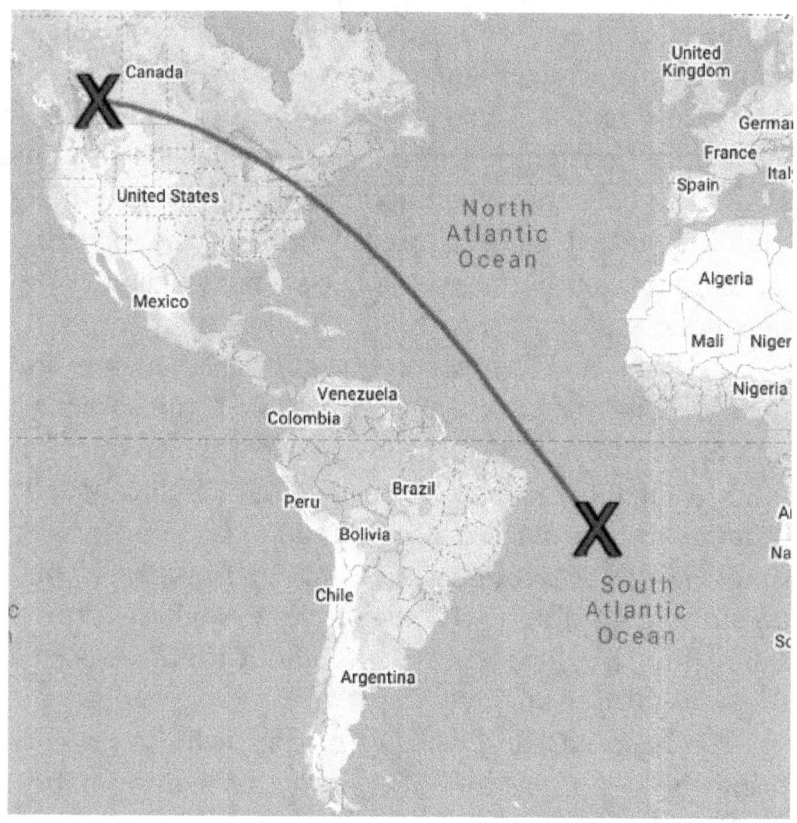

Figure 31. The ground track of the Great Meteor Procession has been traced for more than 7000 miles (11000 km).

Chapter 6
Little Moons

We have seen that John O'Keefe argued against Wylie's meteor shower. Instead, he returned to Chant's hypothesis that the meteors comprising the Great Meteor Procession moved in orbits resembling those of satellites, or at least satellites re-entering the atmosphere. But how could such satellite orbits come to be?

O'Keefe (1961) imagined that it all began when a small asteroid, passing near the Earth, was captured by Earth's gravity. It went into orbit about our planet becoming, for a time, a miniature moon. We begin, then, by asking a question: Does Earth really have natural satellites other than the Moon?

In the 21st century, the search for natural Earth satellites is made difficult by a confounding array of artificial satellites. A huge number of orbiting satellites is rapidly increasing. In contrast, before 1957 orbital space was pristine. Rockets were sending payloads higher and higher, but the various national space programs had launched no satellites yet. Up to 1957, any search for small natural moons would not be confused by artificial ones. By the 1950s, it had, however, become apparent that it would not be long before rockets lofted objects into space with the 5 miles per second speed required to place them into orbit. It is at this point that Clyde Tombaugh (1906 – 1997) enters our story. In the early 1950s, Tombaugh inaugurated and directed a photographic search for natural Earth satellites.

Figure 32. A young Clyde Tombaugh stands beside his homemade telescope around 1928. Wikimedia. Public domain.

Tombaugh is, of course best known for discovering Pluto in 1930. While still an amateur astronomer who had not yet attended college, Tombaugh was called to the Lowell Observatory to carry out a photographic search for a planet beyond Neptune. He was later involved with missile tracking at White Sands, New Mexico, and with the organization of programs for photographing planets.

Great Meteor Procession

Tombaugh explained the rationale behind his search, which would be sponsored by the U.S. Army Office of Ordnance Research:

Paradoxically, the regions of space nearest the Earth were virgin fields for exploration. The progress of developments in rocketry encouraged the hope that man might actually venture into space within the next generation. The senior author was in charge of the optical instrumentation at the White Sands Proving Ground when the two stage combination of the German V-2 and Wac Corporal (known as "Bumper") was fired to the record height of 250 miles. This event was surely a portent of things to come. Multiple stage rockets would be fired higher and higher, until they could be made to orbit around the Earth as satellites. To obtain ballistic data on such missiles would require instrumentation beyond the means and methods then available. To prepare for such a technological advance, the senior author strongly felt that the regions of space around the Earth should be explored in order to ascertain some of the hazards such missiles might experience, and particularly to prevent misidentification with possible natural satellites. The equipment and methods developed for such an exploration would serve to instrument the tracking of artificial satellites, whenever they should become a reality.

The mini-moons Tombaugh sought would be effectively above the atmosphere, meaning that they would shine not as meteors but only by reflected sunlight. Having given careful thought to the technical problems associated with detecting such moving points of light, Tombaugh devised a photographic search program. Northern observations commenced in December, 1953, from the grounds of the Lowell Observatory in Flagstaff, Arizona. Observations were also obtained near the equator, at Quito, Ecuador. Tombaugh and colleagues took advantage of a 1956 lunar eclipse to search for small objects orbiting the moon. The results of Tombaugh's searches would be published in 1957 and 1959. No satellites were found.

Figure 33. Tombaugh's northern search instruments in Flagstaff, Arizona. From Tombaugh (1959). Hathitrust Digital Library.

The searches Tombaugh organized had limits. Not all possible orbits could be scrutinized, and only objects brighter than certain limits could be detected. The technical report for the search for Earth satellites summarized the results:

> *Several fairly promising, but faint, satellite suspects were encountered. Vigorous attempts were made to recover them, but without success. It was concluded that these objects were probably spurious, although they could have been satellites of high eccentricity beyond the practicality of recovery. Indeed, they may have been very small asteroids brushing by the Earth in their elliptical orbits around the sun. The extensive search conducted by this project indicates that the satellite regions of the Earth are remarkably free of natural discrete bodies. It appears that the Earth has only one natural satellite; namely, the Moon.*

Thirty years after Tombaugh's search, Michael Longo and Robert Morris (1986) carried out another search for natural Earth satellites, this time using radar data from the North American Aerospace Defense Command. Despite complications from artificial satellites, no objects likely to be natural mini-moons were identified to heights of 10,000 kilometers. Unlike the Tombaugh survey, the radar search was able to cover all orbital orientations. While Tombaugh's optical search and the later radar survey identified no natural satellites other than the Moon, that negative result subsequently changed.

Theorists recently calculated how often small asteroids are expected to be captured by our planet (i.e., Granvik et al. 2012). They found that at any given time there ought to be at least one small asteroid of diameter 1 meter or more caught and gravitationally bound to Earth. Such mini-moons would not, however, ordinarily be expected to circle Earth in low orbits. The hold of Earth on such mini-moons is likely to be weak and temporary, and they would usually be found far from our planet's surface, far too high for them to glow as meteors.

At least two such objects have been discovered in recent years. Both were very small asteroids. Kwiatkowski et al. (2009) concluded that the 3-meter (10-foot) asteroid 2006 RH_{120} was in orbit about the Earth for a year. It did not get closer to Earth than about 170,000 miles (270,000 km). Fedorets et al. (2020) found that the 1-m asteroid 2020 CD_3 might have remained in orbit about the Earth for several years. Its closest approach to the center of the Earth was about 8000 miles (13000 km).

Gravitational tugs on the motions of these temporary satellites by Moon, Sun, and planets eventually causes most of them to escape again. There is no chance that either of the two aforementioned mini-moons will produce something like the Great Meteor Procession anytime soon. However, close encounters with captured mini-moons can occur. It has been calculated that 2020 CD_3 has a greater than 1 percent chance of hitting the Earth sometime after 2061 and before 2120. Granvik et al. (2012) and Fedorets et al. (2017) estimated that perhaps 0.1 % of all meteors may have been temporarily captured into orbit by the Earth before they hit our planet. With that in mind, let us return to O'Keefe's scenario for the creation of the Great Meteor Procession.

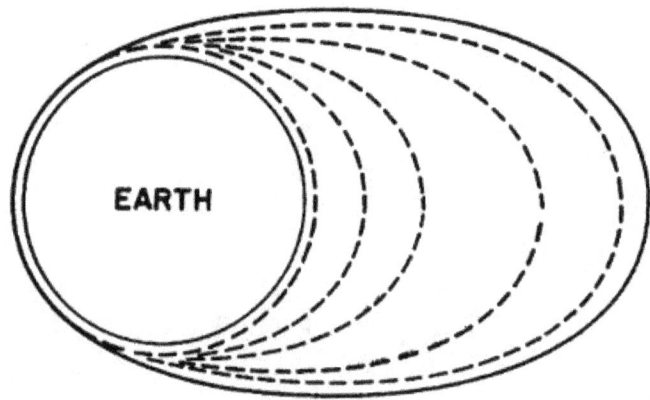

Figure 34. This figure from O'Keefe (1961) shows the orbit of the parent body of the 1913 meteors (solid line) and the orbits of various daughter fragments (dashed lines). The daughter fragments, dipping into the atmosphere, would give rise to the meteors of the procession. Courtesy AAAS.

O'Keefe proposed that an object captured by Earth's gravity ended up in an eccentric orbit which, at its lowest point, came near to the Earth (Figure 34). This captured mini-moon was the parent body of the meteor procession. The parent body began to break apart at the low point in its orbit, with the still orbiting fragments later falling deeper into the atmosphere to produce the meteor procession, perhaps on the next orbit. "If the main body had a small eccentricity, say 0.02[16], then the daughter bodies could have a range in eccentricity from this value to zero...With this choice of eccentricity, the range in [orbital] period is about 3.3 minutes, a figure in agreement with the statements of witnesses that the shower took about this long to pass a given point." O'Keefe thus argued that multiple fragments broken from the parent body, each on a slightly different orbit, combined to make the Great Meteor Procession.

Recently, Martin Beech and Mark Comte (2018) took a fresh look at how incoming meteoroids might produce meteors with long Earth-grazing trajectories. Carrying out computer simulations, they modeled meteoroids plunging into the atmosphere at different initial angles and with speeds of 12 km per second – a speed they regarded as appropriate for an encounter with a captured mini-moon. Beech and Comte found that, under the right conditions, meteors with long, Earth-skimming, trajectories could be produced. Such satellitic objects had heights of 50 ± 10 kilometers (31 ± 6 miles) and speeds reduced to 8 km per second (5 miles per second). They would typically travel about 1600 km (a thousand miles) until they eventually succumbed to air resistance. Thus, once again we see that, in this scenario, no single meteor was likely to have been seen over the entire length of the 1913 procession. Too steep an encounter angle would shorten the track of a meteor, sending it into the ground. Too shallow an entry angle

---

[16] A circular orbit has an eccentricity of 0. An eccentricity of 0.02 means that at the satellite's farthest distance from the center of the Earth it will be 1.04 times its nearest distance — so not as eccentric as the parent body orbit drawn in Figure 34.

Great Meteor Procession

would allow the Earth-grazing meteoroids to dip into the atmosphere, only to exit and re-enter again at some later time.

Beech and Comte estimated the parent mini-moon needed to produce the 1913 procession might have been 3 or 4-meters in size, the size of an automobile, with a mass of 75000 kg, some 80 US tons, although considerable uncertainty attached to those numbers. The total energy deposited in the atmosphere by the meteoroids was estimated to have approached tens of kilotons – equaling or exceeding the energy release of the first atomic bombs.

In a fragmenting mini-moon with an Earth-grazing trajectory, we thus have a plausible physical picture for the production of the Great Meteor Procession. In the next two chapters, we turn back to the historical record. Was the event of 1913 really the only display deserving the name Great Meteor Procession?

Great Meteor Procession

Chapter 7
Earth-Grazers

In the years since 1957, re-entering artificial satellites have duplicated aspects of the 1913 fireball procession. When the Mir space station deliberately re-entered in March 2001, it broke apart as it plunged deeper into the atmosphere, its fragments becoming meteoric in appearance. Pieces decreased in altitude from 60 miles (100 km) to impact in the South Pacific in a timespan of about 16 minutes. During that interval, Mir traveled about 5600 miles (9000 km), a distance approaching the observed distance traversed by the 1913 procession. Re-entering Mir did not, however, produce the successive fireballs seen in 1913, so the match to the 1913 procession was by no means perfect. However, let us ignore artificial satellites and restrict ourselves to natural meteors. Historical accounts identify some meteors that might have been in Earth-grazing trajectories. We next compare some of them to the Great Meteor Procession.

An early well-documented account of a possible grazing fireball dates to the summer of 1783, shortly before the Treaty of Paris recognized an independent United States. At that time, the cause of meteors had not yet been established, but fireballs were garnering increased scientific attention. Newspapers in Scotland and England reported fireballs streaking overhead on the evening of August 18. A detailed description by Tiberius Cavallo, a visiting Italian who viewed from the terrace of

Great Meteor Procession

Windsor Castle was published in the *Philosophical Transactions of the Royal Society*:

> Some flashes of lambent light, much like the aurora borealis, were first observed on the northern part of the heavens, which were soon perceived to proceed from a roundish luminous body, whose apparent diameter equaled half that of the moon, and almost stationary in the same point of the heavens (see A in the annexed figure...). It was then about 25 minutes after nine o'clock in the evening. This ball, at the beginning, appeared of a faint bluish light, perhaps from appearing just kindled, or from its appearing through the haziness; but it gradually increased its light, and soon began to move, at first ascending above the horizon in an oblique direction towards the east. Its course in this direction was very short, perhaps of five or six degrees; after which it directed its course towards the east ... its light was prodigious. Every object appeared very distinct; the whole face of the country, in that beautiful prospect before the terrace, being instantly illuminated. At this moment the body of the meteor appeared of an oblong form like that represented at B in the figure; but it presently acquired a tail, and soon after it parted into several small bodies each having a tail and all moving in the same direction, at a small distance from each other, and very little behind the principal body the size of which was gradually reduced after the division (see D in the figure).

Great Meteor Procession

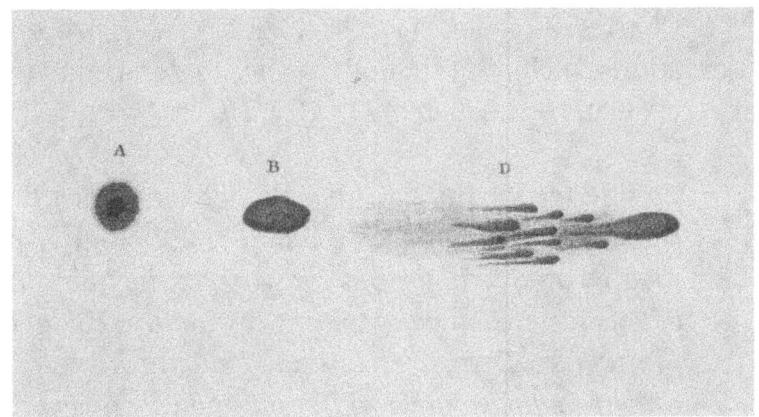

Figure 35. Three different stages in the appearance of the 1783 fireball, as depicted in Cavallo's paper in the Philosophical Transactions of the Royal Society.

Figure 36. A print by Thomas and Paul Sandby showing the Meteor of August 18, 1783, as seen from the East Angle of the North Terrace, Windsor Castle. Wikimedia. Public domain.

Charles Blagden (1748 – 1820) used several observations, including that of Cavallo, to determine the path traversed by the meteors (1784). He wrote

> *To adduce the different accounts from which this path is determined', would not only be insufferably tedious, but contrary to the intention of this letter, which is to give a summary view of the whole. They are contained partly in letters, and partly in the different news-papers of England and Scotland, most of which have been perused for this purpose. The information derived from the news-papers, however incorrect in the detail, is brought to some degree of certainty by the check of comparing, them with one another.*

He found that the meteor had crossed over Scotland from the North Sea, then traveled to the southeast across eastern England before it proceeded over the English Channel. Beech's (1989) updated version of Blagden's path gets rid of a small change in direction Blagden thought occurred near Yorkshire, but is generally similar. The track from the North Sea to the English Channel continued to France and, doubtfully, as far as Italy. The meteors were seen over a path stretching at least a thousand miles (1600 km), placing the 1783 fireball in the category of grazing meteors.

Two centuries later, on August 10, 1972, vacationers in the western United States were astonished to see a brilliant daytime fireball. Observations from the ground and from space-based sensors, showed that the dazzling visitor penetrated the atmosphere over Utah, descended to a minimum height of about 36 miles (58 km), and then increased in altitude to escape the atmosphere over Alberta, Canada, flying off into space once more (Z. Ceplecha, 1994; R.D. Rawcliffe et al., 1974). The mass of the 1972 meteoroid has been estimated to be between a hundred and a thousand metric tons. Its Earth-grazing path traversed some 940 miles (1500 km), long but much less than that of the 1913 procession.

The 1783 and 1972 fireballs were particularly dramatic events. Grazing fireballs in the broad sense do not, however, appear to be extremely rare. Half a century after the 1972 fireball, Shober et al. (2020) published a paper in the *Astronomical Journal* titled "Where Did They Come From, Where Did They Go: Grazing Fireballs". They listed nine instances of possibly grazing meteors observed between 1990 and 2014, and made a particular study of a meteor observed in 2017 by the Desert Fireball Network, an array of automated cameras set up to watch Australian skies.

On the night of July 7, 2017, the cameras caught a fireball streaking overhead. The fireball grazed the atmosphere, entering at a shallow 4.6-degree angle. The 14-92 kg (30 – 200 pound) meteoroid began to glow at an altitude of 85 km (53 miles), plunging to a height of 58 km (36 miles). However, the fireball then skipped back up, remaining visible until it had climbed to 86 km. At closest approach to the ground, atmospheric resistance caused a fragment to break off the main meteoric body. The fragment followed a path through the sky similar but not identical to that of the main body. The glowing trail of this fragment could be seen for only a small part of the distance traveled by the main meteor. One can imagine similar fragments might have broken from the meteors of the 1913 procession.

During the time that it was seen, the Australian fireball traveled some 1300 km (800 miles), a distance comparable to the 1972 bolide, but, once again, far less than the distance crossed by the 1913 procession. When first recorded, the meteoroid was traveling at 15.7 km/s. It had slowed only slightly, to 14.2 km/s when it faded from view. Recall that the orbital speed of a low-Earth-orbit satellite is about 7.8 km/s, only half the speed of the meteoroid in this case. The 2017 meteoroid thus retained more than enough energy to escape the Earth (escape speed 11.2 km/s).

Shober et al. went on to calculate the orbit around the Sun of the meteoroid responsible for the 2017 fireball before and after its encounter with Earth. Before its brief flight as a meteor, the meteoroid was in an orbit that carried it near the Earth at perihelion (closest approach to the Sun). When farthest from the Sun, at aphelion, it was beyond the orbit of Mars, but not quite to the orbit of Jupiter. The encounter with Earth altered its orbit. After its encounter, the meteoroid goes to the orbit of Jupiter at aphelion. Its orbit will become less certain in the future as Jupiter's gravity repeatedly disturbs its motion.

These reports show that the procession of February 9, 1913 is not in all respects a one-of-a-kind event. Grazing meteors with paths stretching about a thousand miles through the atmosphere are rare but not extremely rare. Nonetheless, the daylight fireball of 1972, the Australian fireball of 2017, and even the British fireball of 1783 did not equal the Great Meteor Procession in number of meteors or length of trajectory. However, half

Great Meteor Procession

a century before the 1913 procession, there occurred a meteoric display which merits special attention and comparison with the Great Meteor Procession. That is the subject of our next chapter.

Great Meteor Procession

Chapter 8
Year of Meteors

Figure 37. Walt Whitman photographed at Matthew Brady's studio, c. 1862. Public domain.

Great Meteor Procession

When Walt Whitman published a new edition of *Leaves of Grass* in 1867, one of its poems was *Year of Meteors (1859-1860)*:

Year of meteors! brooding year!
I would bind in words retrospective some of your deeds and signs,
I would sing your contest for the 19th Presidentiad,
I would sing how an old man, tall, with white hair, mounted the scaffold in Virginia,
(I was at hand, silent I stood with teeth shut close, I watch'd,
I stood very near you old man when cool and indifferent, but trembling with age and your unheal'd wounds you mounted the scaffold;)
I would sing in my copious song your census returns of the States,
The tables of population and products, I would sing of your ships and their cargoes,
The proud black ships of Manhattan arriving, some fill'd with immigrants, some from the isthmus with cargoes of gold,
Songs thereof would I sing, to all that hitherward comes would I welcome give,
And you would I sing, fair stripling! welcome to you from me, young prince of England!
(Remember you surging Manhattan's crowds as you pass'd with your cortege of nobles?
There in the crowds stood I, and singled you out with attachment;)
Nor forget I to sing of the wonder, the ship as she

Great Meteor Procession

swam up my bay,
Well-shaped and stately the Great Eastern swam up my bay, she was 600 feet long,
Her moving swiftly surrounded by myriads of small craft I forget not to sing;
Nor the comet that came unannounced out of the north flaring in heaven,
Nor the strange huge meteor-procession dazzling and clear shooting over our heads,
(A moment, a moment long it sail'd its balls of unEarthly light over our heads,
Then departed, dropt in the night, and was gone;)
Of such, and fitful as they, I sing—with gleams from them would I gleam and patch these chants,
Your chants, O year all mottled with evil and good—year of forebodings!
Year of comets and meteors transient and strange—lo! even here one equally transient and strange!
As I flit through you hastily, soon to fall and be gone, what is this chant,
What am I myself but one of your meteors?

The poem is filled with omens foreshadowing the coming of the Civil War in 1861. Meteor has metaphoric meaning in Whitman's poem, but was there a real meteor inspiring *Year of Meteors*? McBeath et al. (2011) noted four prominent meteors in 1859-60 that might have served as inspiration. Donald Olson (whom we have already encountered in connection with the lengthening of the track of the 1913 meteors) and his colleagues specified one spectacular 1860 meteoric display as the inspiration for Whitman (Olson et al. 2010).

Figure 38. Report on the 1860 meteors from the New York Daily Herald for July 23, 1860. I am impressed by how many specific details about the meteors were often printed in the newspapers of 1860.

On the evening of July 20, 1860, those outdoors in the northeastern United States had their attention suddenly pulled skyward. Unlike the winter meteors of 1913, the meteors of 1860 appeared on a warm summer evening, when many people were outdoors at the end of the day. The first recorded observations of the eastward-traveling meteors came from eastern Michigan. No reports came from Wisconsin, further to the west, but clouds hindered observations there. From Michigan, the meteors were seen over Ontario, New York state, and Pennsylvania, crossing over Long Island before vanishing in the distance as they moved out to sea. A letter from Harvard astronomer George P. Bond (1825 – 1865) in the *New York Tribune* for August 2 gave a concise account:

> *The following is an outline of the path described by the great meteorite[17] of the 20th inst., as derived from observations made in various sections of the country: It first became visible in the region of the great lakes, pursuing a course E. S. E. from the northern part of Michigan in a straight line over Lake Huron, Canada West, Lake Erie, South-Western New-York, the North Eastern part of Pennsylvania. South-Eastern New-York. the south-west corner of Connecticut, Long Island Sound, and Long Island. It was seen 300 or 400 miles out at sea, and probably passed on to resume its path in the solar system, undoubtedly a good deal disturbed from its original orbit by the earth's attraction and the resistance offered by its atmosphere.*
>
> *Its nearest approach to our globe was within 35 miles; it was at this elevation, nearly, when crossing the Hudson, 15 miles in an air-line*

---

[17] In current usage, what was seen was a meteor, not a meteorite.

Great Meteor Procession

from New-York. Its velocity was about 20 miles a second. The vertical diameter was one-fourth of a mile, including the brightest portion of the luminous atmosphere surrounding the nucleus.

    Bond's letter went on to complain about something we have already seen seconded by both Chant and Wylie. Not all observer reports were equally good. Bond found it impossible to reconcile all of the observations he received, requiring him to distinguish the better from the poorer reports. Though obviously interested in the 1860 meteors, Bond does not seem to have published a formal paper on the subject. Suffering from tuberculosis, and struggling to keep the Harvard College Observatory in operation during the Civil War, he would die early in 1865.

    Harper's Weekly was one of the largest circulation magazines in the United States as the Civil War approached. Two days after Bond's letter appeared in the *New York Tribune*, *Harper's Weekly* gave front page coverage to the meteors in its August 4, 1860 issue. Vivid eyewitness descriptions of the meteors were gathered and printed. However, the most arresting aspect of the *Harper's* article is its illustrations. Engravings of the meteors were published which were based upon sketches drawn by eyewitnesses.

Great Meteor Procession

THE METEOR OF JULY 20, AS SEEN BY J. A. ADAMS, ESQ., AT SARATOGA SPRINGS.

Figure 39. An engraving in Harper's Weekly from a sketch of the meteors made at Saratoga Springs, north of Albany, New York.

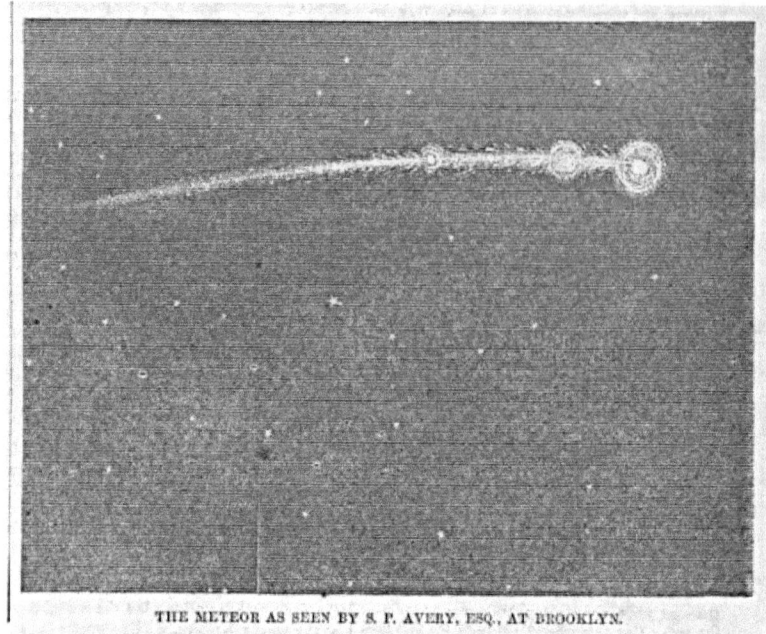

Figure 40. This Harper's Weekly engraving was based upon a sketch made in Brooklyn, about 164 miles (264 km) south of Saratoga Springs. Both Figures 39 and Figure 40 show two main fireballs, although the Brooklyn engraving also shows a lesser fireball in the train.

Figure 41. The third Harper's Weekly engraving, depicts the display as it appeared near Bedford on Long Island, It, too, shows two main fireballs. A number of lesser meteors follow.

Frederic Church (1826 – 1900) was a major artist of the Hudson River School. By 1860, he had gained a reputation for dramatic landscapes. On the evening of July 20, Church witnessed the fireballs over the Catskills. He subsequently painted a striking picture of the meteors in flight. It shows two principal fireballs, followed by a luminous train containing smaller points of light. Church's painting helped lead Olson et al. (2010) to their conclusion that these were the meteors important to Whitman's *Year of Meteors 1859-1860*.

Figure 42. A grayscale version of Frederic Church's painting of the 1860 meteors. Public domain.

Later in 1860, Yale astronomer Chester S. Lyman (1814 – 1890) contributed a quick study of the meteors to the *American Journal of Science and Arts* (1860). However, at first no paper appeared analyzing the July, 1860, meteors with the depth of Chant's paper on the 1913 event. That changed when James H. Coffin (1806-1873) took up their scientific study.

Coffin is known to meteorologists for his pioneering studies of winds. However, as professor of mathematics and astronomy at Lafayette College in Pennsylvania, his interest extended beyond the weather. Coffin gathered observations of the meteors from newspapers and eyewitnesses. He personally interviewed a number of witnesses in Pennsylvania, employing a theodolite[18] to reconstruct the direction and apparent altitude of the fireballs. Coffin noted a circumstance which aided his efforts: He credited the large number of observers of the meteors to "the prevalent custom of our people, to sit at the front doors of their houses in summer evenings."

After the Civil War, Coffin published his findings in the *Smithsonian Contributions to Knowledge* (1869). The paragraphs below summarize what he found and are generally consistent with Bond's early letter:

> On *the evening of July 20th, 1860, a meteoric fireball passed over the northern parts of the United States and the adjacent parts of Canada, of so extraordinary brilliancy as to attract the attention of numerous observers along its entire visible track of nearly or quite 1300 miles, and on either side of it to the distance of several hundred miles. It was reported to have been first seen moving eastward from a point nearly over the western shore of Lake Michigan, though it not improbably became luminous when it was somewhat further west, as the sky over all that region was obscured by clouds, and it was not till the meteor had reached a point some 150 miles further east that the first reliable determination of its position was made. From thence many eyes watched its course till it disappeared quite out at*

---

[18] A surveying instrument.

> sea in a southeasterly direction from the island of Nantucket.
>
> From the following series of observations, obtained partly from the newspapers of the time, partly through the co-operation of scientific friends, who, at the request of the writer, kindly made inquiries in regard to the phenomenon in their respective localities, or measurements of the meteor's position, as estimated by themselves or pointed out by those who saw it; and partly from collections kindly put into his hands for the purpose by the Smithsonian Institution, and by Profs. Lyman and Newton, of Yale College, an attempt has been made to determine the elements of its orbit or path.

Anticipating Chant's study of the 1913 fireballs, Coffin published eyewitness reports of the 1860 fireballs as they were seen from various locations. The track of the fireballs he derived (Figure 43) is strikingly similar, although not identical, to a portion of the track which Chant would derive half a century later for the 1913 meteor procession.

To better appreciate the 1860 meteors, we will look at a couple of the accounts included in Coffin's paper. The *Buffalo* (New York) *Courier* for July 21, 1860, reported:

Great Meteor Procession

Last night, about half past nine[19], the grandest meteor we ever had the fortune to see, made its way through the heavens to the wonderment of every mortal with eyesight who was out of doors at the time. It sprang into view, as near as we could ascertain, at or near the horizon almost exactly in the west. We were standing, at the moment, in the shadow of buildings that completely shut out the western sky. A flood of light, like that of a vivid, continuous flash of lightning, or like a bright dawn, streamed over the tops of the houses, and grew in intensity for a few seconds, ere the majestic orb sailed sublimely into sight overhead. Over the zenith it sped, reddish in hue, and with a wake of fire that spanned the sky for an instant like a vast arch of celestial flame.

The *Pottsville Mining Journal* for July 21 reported on the appearance of the meteor as seen from Danville, Pennsylvania:

A very brilliant meteor passed over this place last evening at ten o'clock, giving as much light as a full moon. It came in view at the horizon, west of northwest, and passed due east, being about six seconds[20] in passing. It went out of sight below the horizon, east of northeast. When

---

[19] This was before the implementation of standard time zones. Times reported to Coffin would have been the local times of the observers or the time in Washington or some other prominent city.

[20] One of the shortest duration estimates among the reports.

Great Meteor Procession

directly northeast, it broke, forming two, one following the other. Some minutes (another account says four) after it disappeared, a sound resembling thunder was distinctly heard. No clouds were in sight.

Reports of associated sounds came from witnesses in parts of Pennsylvania and New York. From Rochester, New York we learn that: *"A report was heard here about three minutes after the meteor passed out of sight-for it seemed to explode as it disappeared, and then appeared double."* Dr. E. Joerg in Coudersport, Pennsylvania, heard a "report" five minutes after the passage of the meteor. Surveyor Wilson King in Erie, Pennsylvania, heard an "explosion" three to five minutes after the meteor passed. Explosions were also heard at Matteawan (New York), along the Hudson River, and at Branford on the Connecticut coast. However, other witnesses along the same portion of the meteor track saw the luminous display but made no mention of subsequent noises.

Coffin devoted great effort to determining the ground track of the fireball, and the map included in his paper is shown in Figure 43. Coffin also calculated the height of the meteor above the ground during the course of its observed flight (Figure 44). He expended great effort on securing accurate reports of where in the sky the meteor was seen, paying particular attention to the apparent location of the meteor against the background of the stars. While he used theodolite measurements to obtain the direction and apparent altitude of the meteor, those had to be obtained after its flight and so depended on the accuracy of memories. It is no wonder that he also

Great Meteor Procession

found it difficult to reconcile all reports. His reconstruction of the height profile put the minimum altitude of the meteor near 39 miles when it was approaching the New York City area.

How does the great meteor of 1860 stack up compared to the great meteor procession of 1913? Unlike the procession of 1913, in 1860 fireball did not follow fireball. In 1860, a single brilliant fireball with a following train of light broke into two bright fireballs followed by a string of lesser lights. The 1860 meteor does, however, fall into the class of Earth-grazing fireballs. Although the path of the 1860 meteors across the United States and Canada is remarkably like that of the 1913 procession, that would appear to be coincidental. There is no reason to expect an incoming meteoroid on July, 1860 to have a connection with those of February, 1913. If Coffin's analysis is close to being correct, in 1860 the meteors were higher both at the beginning and end of their observed path than in between, while their speed may have exceeded Earth's escape speed. We thus have the possibility that the meteoroid escaped again into space, as did the daylight fireball of 1972.

The observed track of the 1860 fireball was 1300 miles (2090 km), and could in actuality have been longer. That, however, is not close to the 7000 miles of the Great Meteor Procession. The meteors of the *Year of Meteors* were no duplicate of the Great Procession – but, on the other hand, impressive as the Great Meteor Procession was, it did not inspire a memorable poem.

In the next chapter, we turn our attention from meteors to the strange realm of unidentified flying objects.

Great Meteor Procession

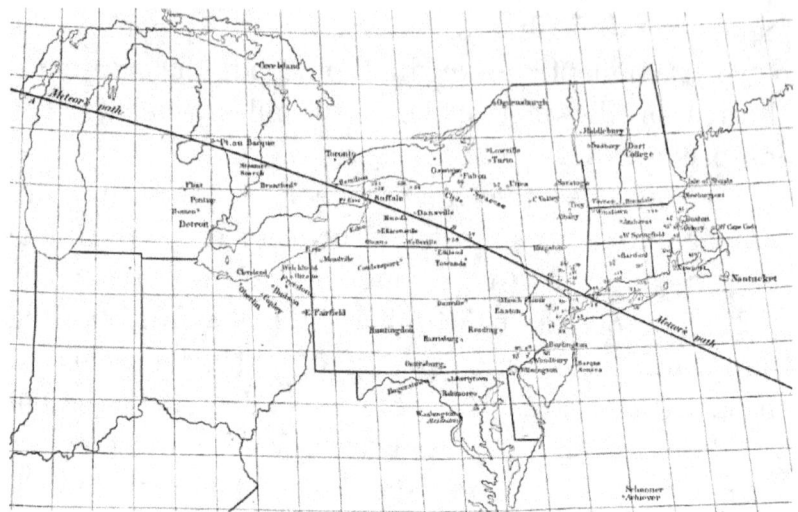

Figure 43. The path of the fireballs of July 20, 1860, according to James H. Coffin.

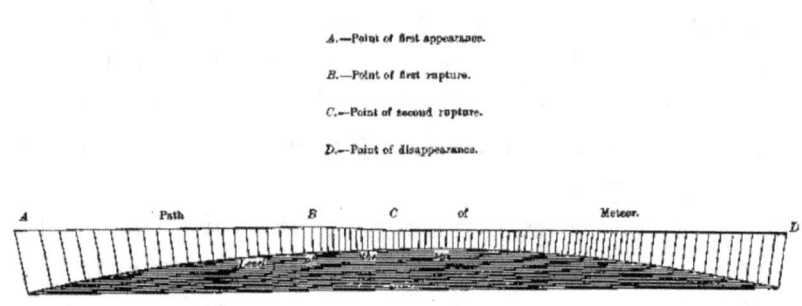

Figure 44. The profile of the altitude of the meteor according to Coffin.

Chapter 9
Meteors or Aliens?

While some witnesses of the 1913 meteors likened them to maneuvering airships, they probably did not mean that literally. Charles Fort, in his 1923 book *New Lands,* was perhaps the first to intentionally suggest in print that the fireballs of February 9, 1913 were not a natural phenomenon. However, as described in chapter 3, Fort never offered an opinion as to exactly what the fireballs were.

Fort, of course, wrote *New Lands* long before flying saucers were thrust into the popular press by Kenneth Arnold's June 24, 1947 report of unidentified flying objects over the state of Washington[21]. The quickly rescinded announcement of the recovery of a crashed disk at Roswell, New Mexico, in early July, kept UFOs in the news, although the Roswell incident itself would be largely forgotten until the late 1970s. As UFOs continued to be reported, the notion grew among some that they were spaceships piloted by visiting extraterrestrial beings. This extraterrestrial hypothesis was widely debated before the 1940s ended.

Newspapers, magazines, books, films, radio, and television all spread the saucer story as the 1940s turned

---

[21] In the late 1940s and during much of the 1950s, mysterious objects in the sky were often termed flying saucers. Later, unidentified flying objects became the more common name, joined, still more recently, by unidentified aerial phenomena.

into the 1950s (Bertram 2006). In the late 1940s, Raymond Palmer's science fiction magazine, *Amazing Stories,* promoted the existence of extraterrestrials, while the pages of *Fate,* a magazine co-founded by Palmer in 1948, carried accounts of alien visitors. Two popular books published in 1950 helped keep flying saucers in the public mind. Frank Scully's *Behind the Flying Saucers* told of the recovery of alien bodies from downed flying saucers and made the *New York Times* best-sellers list. Donald Keyhoe's *The Flying Saucers are Real* informed readers that the military was hiding the true story of flying saucers. Films also played a part in setting-up a mindset in which extraterrestrial visitations were, for better or worse, possible. Flying saucers appeared in movies such as *The Thing from Another World* (1951) and *The Day the Earth Stood Still* (1951).

Within days of Arnold's sighting and just before the Roswell incident made headlines, the Pittsburgh *Sun-Telegraph* for July 7, 1947, carried an article by E.E. Lewis which attributed to Charles Fort the idea that the meteors of February 9, 1913 were spaceships. Lewis wasn't quite right on that point, but the appearance of the *Sun-Telegraph* article indicates that Fort's account of the meteor procession had resurfaced with the coming of the flying saucers. From 1947 on, Fort's eccentric ideas would be enmeshed with the story of UFOs.

It is worth digressing to clarify that sightings of unknown aerial objects were not new in 1947 or even in 1913. Mystery airships had filled newspaper columns well before the year of the Great Meteor Procession. In 1896-1897 reports spread across the United States of unidentified airships overhead, while in 1909 a supposed

Great Meteor Procession

airship surprised Christmas shoppers in New England. The years 1912-1913 saw sightings of unidentified airships over Great Britain. These mystery airships were often credited to secretive inventors, but on rare occasions a daring writer ventured an extraterrestrial origin. In some cases, an extraterrestrial origin could be verified, as when in 1909 a young H. P. Lovecraft, today known for his horror and fantasy fiction, informed gawkers on a Providence street-corner that what they took to be an airship's searchlight was in fact the planet Venus! However, before the flying saucer era, the 1913 fireballs, if they were remembered at all, appear to have been regarded as natural, albeit unusual, objects, except perhaps by Fort's admirers.

After 1947, the situation changed. Natural explanations for the Great Meteor Procession were disputed by those who believed that the fireballs were spaceships from beyond our planet. In the remainder of this chapter, we take a look at the connection between UFOs and the Great Meteor Procession, as it developed in the 1950s and 1960s.

It is important to note that not all popular retellings of the story of the 1913 fireballs took the spaceship path. For example, Willy Ley, refugee from Hitler's Germany and author of books on rocketry, history, and strange animals, wrote about the meteor procession in a 1951 article in *Galaxy*, one of the post-war science fiction pulp magazines. Ley stuck close to the Chant explanation. Titled *The Meteoric Stream*, his article carried the subheading "Like cops, astronomers usually aren't around when big things happen – which may explain reports of Flying Saucers!" Ley ended his account not with spaceships but with a different worrisome thought: "with world tension as high as it is, nine out of ten observers [in 1951] would interpret the procession as a military attack."

Jacques Vallée, in his UFO book *Anatomy of a Phenomenon* (1965), suggested that the Canadian fireballs of 1913 might have had some connection with "airships" seen over Britain early in that year, but he did not venture anything more. However, others were straightforward in proclaiming that the fireballs were piloted craft. For illustrative purposes, I limit our discussion to three well-known writers on UFOs, each of whom advanced a non-natural explanation for the fireball procession.

Writer and radio and television show host Frank Edwards (1908 – 1967) is perhaps best remembered for his 1966 book *Flying Saucers – Serious Business*, which was on the *New York Times* best-sellers list for sixteen weeks. The Great Meteor Procession is not mentioned in that book. However, before *Flying Saucers – Serious Business*, Edwards contributed the *Great Sky Procession of 1913* to *Fate* magazine for December, 1961. He also included the procession in his 1964 book *Strange World*. Edwards dismissed the natural satellite explanation for the fireballs, concluding that "the simplest explanation which covers all the known facts in this case — and the only conclusion that does cover all the facts — is that these things were under intelligent direction."

Brad Steiger (1936 – 2018) was a prolific author of books on paranormal topics. In his 1966 book *Strangers from the Skies*, Steiger noted that in "1913, it would have been most startling to suggest that the objects might not have been meteors at all but intelligently manned space craft which, far from disintegrating over the northeast tip of Brazil, had made the decision to return to outer space after a partial orbit of the Earth."

John Keel (1930 – 2009) is perhaps best known as the author of the 1975 book *The Mothman Prophecies*, upon which the 2002 movie of the same name was based. In an earlier book, *Operation Trojan Horse* (1970), Keel called 1913 a meteor flap year, following the "UFO flap" terminology for a year especially rich in UFO reports. Like Fort, Keel did not fully specify what the fireballs were, but he wrote that their slow, majestic motions were unlike "natural meteors". Keel, however, by then had begun to doubt the alien spaceship explanation for UFOs, arguing instead that they were directed by "ultraterrestrials," perhaps some sort of supernatural beings native to Earth.

All three approached the story of the 1913 procession in somewhat similar fashion. To them, the Great Meteor Procession was interesting, but of secondary interest compared to the contemporary UFO reports which were their main staple. The discussions by Edwards are the most extensive, and even his chapter on the subject in *Strange World*, "Express Train in the Sky", is only eight pages out of a 251-page book (paperback edition). None of the three delved deeply into the events of 1913, none sought to find new information about the meteors, none presented arguments for the artificial nature of the fireballs other than their unusual grouping, long flight, and slow, "majestic" motion.

Post-war UFOs also caught the attention of more scientific students of meteors. Interestingly, some investigators whom we met in previous chapters also became involved in the study of UFOs. Lincoln LaPaz, the scientist who encouraged Mebane's investigations, is one such person.

Figure 45. Lincoln LaPaz directs the recovery of a 2,360 pound (1070 kg) meteorite which fell in 1948 in Furnas County, Nebraska. Courtesy of the University of New Mexico Archives.

Great Meteor Procession

Project Sign, begun in 1948, was an early program of the U.S. Air Force tasked with evaluating UFOs. Meteor expert LaPaz, who had worked at the New Mexico Proving Ground during World War II, was approached by Sign to consult on a Memphis, Tennessee UFO sighting, a case which also saw astronomer J. Allen Hynek's first involvement with UFOs (Swords et al. 2012). When, in late 1948 and 1949, green fireballs were seen over New Mexico, LaPaz, who held a security clearance, was asked to look into the matter. LaPaz concluded that not all of the fireballs could be explained as natural meteors, but as we shall see below, he failed to convince other scientists of his position. Their green color, he thought, might indicate the presence of the element copper[22] within the meteors, perhaps a sign of artificial (but probably not unworldly) construction. It is not improbable that the green fireballs encouraged what may have been an existing interest in the Great Meteor Procession[23].

Another important investigator of the 1913 meteors, Alexander Mebane, was also intrigued by UFOs, and he became involved with their investigation during the 1950s. In 1954, Mebane was a founding member of the Civilian Saucer Intelligence (CSI) group in New York, at the same time as he was carrying out his search for undiscovered observations of the Great Meteor Procession (chapter 4). How Mebane, a chemist, became

---

[22] The green color of some meteors is not usually attributed to copper today. Nickel can produce a green color. Magnesium is also said to make a meteor green, but gives a bluish-green color.

[23] A painting of a "green fireball" by Lincoln's LaPaz's wife, Leota, appeared in the April 7, 1952, issue of *Life* magazine.

interested in meteors is unclear, but that he would also be interested in UFOs is perhaps not surprising. The CSI group was active until 1959 and provided information to the Air Forces' astronomical consultant on UFOs, J. Allen Hynek (Clark 2012). One letter from Mebane to Hynek is dated May 31, 1956, the year which saw the publication of Mebane's paper in *Meteoritics*. Unfortunately for our purposes, the letter deals not with the then four-decade-old meteor procession, but with more contemporary UFO sightings. The letter concludes by noting that Hynek's letters to the CSI were read only by the founders of the group – Mebane, Isabel Davis, and Ted Bloecher.

From March, 1957, until November, 1958, articles with the CSI byline appeared in *Fantastic Universe*, a science fiction magazine (Collins 2019). Mebane was apparently involved with the writing of at least some of the articles, but they usually deal with UFOs of the 1940s and 1950s rather than older reports. I have not come across a mention of the 1913 procession in the issues I have been able to examine, although the books of Charles Fort are mentioned several times. Mebane and the CSI also arranged for publication of an English-language edition of *Flying Saucers and the Straight-Line Mystery* by French UFO researcher Aimé Michel (1958). Mebane added an appendix in which he tried to put Michel's approach on a sounder statistical footing.

While LaPaz and Mebane may have had some sympathy for exotic explanations of UFOs, C. C. Wylie did not. He expanded his skepticism to flying saucers, which he denounced as early as a December, 1947, meeting of the American Association for the Advancement of Science. As noted in chapter 3, his 1953 paper in *Science* was titled *Those Flying Saucers*. Wylie's main purpose for writing that paper was not to dismiss the 1913 meteor procession, but to criticize reports of UFOs. The 1913 meteors were only brought into the paper to illustrate the mistakes that might be made. Wylie described many occasions upon which supposed flying saucers turned out to be ordinary objects. One can hear his irritation in the lines "I have been called out of bed at 1:30 A.M. to explain that the bright light low in the eastern sky is the planet Jupiter." As for the green fireballs that puzzled LaPaz, Wylie was likewise dismissive. He commented that the color green was as common among meteors eighty years ago as it was now.

Harvard astronomer Donald H. Menzel (1901 – 1976), was another prominent skeptic of the extra-terrestrial hypothesis (He was also skeptical of the reality of black holes). Menzel's first book on UFOs, *Flying Saucers*, was published in 1953. Nonetheless, although Wylie's criticisms of the 1913 meteor procession had not at that time been seriously contradicted in print, Menzel chose not to follow Wylie. Instead, he returned to Chant's idea of a meteor procession. He wrote that "Flying-saucer enthusiasts have made much of an unusual meteor display that occurred on 9 February 1913. A great procession of slowly moving meteors moved diagonally across the United States and Canada, from Saskatchewan to Bermuda...It caused, as usual, great consternation among the superstitious[24]. The records clearly show that

---

[24] In fact, superstitious reactions appear to have been rare, although not unknown, in the accounts Chant used. Of course, superstition might have been more common among the general public than those sending letters to Chant.

the objects, which various people estimated at hundreds or thousands, were truly meteoric, though they moved with exceptional slowness." Like Wylie, Menzel scoffed at LaPaz's arguments regarding the green fireballs over New Mexico: "In my opinion, any astronomer who avers that green meteors are new, or that the color must come from burning copper, cannot be much of an authority." LaPaz is not named in Menzel's book, but, since LaPaz had put forward the idea of copper-rich and possibly unnatural green meteors, the implied rebuke is stinging.

Ten years later, Lyle G. Boyd and Menzel published *The World of Flying Saucers: A Scientific Examination of a Major Myth of the Space Age*. The 1913 meteors are also mentioned in that book, but by the time it was written Wylie's explanation had been rejected by Mebane and O'Keefe. Boyd and Menzel accepted the idea that the Great Meteor Procession traversed thousands of miles, but scorned the idea of alien spaceships. They wrote that "But if the UFO cult had existed in 1913, the flying-saucer enthusiasts would probably have regarded the fireball procession as a fleet of spaceships, and would have speculated on the problem of what planet dispatched them and for what purpose."

From 1966 until 1968, the University of Colorado UFO Project was conducted under the overall direction of physicist Edward Condon (1902 – 1974) with funding from the Air Force. Its *Scientific Study of Unidentified Flying Objects* (1968),[25] paid some attention to the 1913 meteor procession, a subject addressed in William Hartmann's chapter on perceptual problems. Hartmann

---

[25] Often called the Condon Report.

likened the 1913 fireballs to the 1968 re-entry of the Zond IV satellite.

Despite their considerable differences, LaPaz, Mebane, Wylie, Menzel, Boyd, and Hartmann all treated the fireballs of 1913 as some type of natural phenomenon. At least they did so in print. If any of them thought otherwise in private, those thoughts stayed well hidden. While the spaceship explanation for the 1913 fireballs repeatedly popped up in popular UFO literature, it made no headway among astronomers.

This chapter is beginning to stray from The Great Meteor Procession, but before it ends, I would like to note that two astronomers who figured in our chapters, Clyde Tombaugh[26] and Lincoln LaPaz, were caught up in a fake-news satellite story in 1954. In the spring of that year, Donald Keyhoe, by then a major author on flying saucers, told reporters that two artificial satellites had been discovered circling the Earth (see, for example, the *Salt Lake Tribune* for May 14). On August 23, 1954, *Aviation Week* announced a scoop: Two satellites of the Earth, probably natural, had been detected, orbiting 400 and 600 miles above the ground. That such satellites existed was, of course, startling because it was more than three years before the launch of Sputnik. In a foreshadowing of what would actually happen in October, 1957, the magazine stated that "the Pentagon thought momentarily that the Russians had beaten the United States to space operations."

---

[26] Tombaugh had himself seen lights in the sky which puzzled him (Swords 2012).

Great Meteor Procession

Aviation Week attributed the discovery to LaPaz, but Tombaugh quickly became caught up in the controversy because of his fame and his role in the Army-sponsored search for small natural Earth satellites (chapter 6). Colonel Walker Holler, head of the Army Office of Ordnance Research at Duke University, made a public statement that did little to quiet things: "if anything is uncovered, depending upon what is uncovered, all or some of it may be classified at the time." Holler's statement, although disclaiming any satellite discoveries by the Army program, was hardly a blanket denial and it kept the pot stirred, although the colonel went on to say that at present the OOR-sponsored hunt for satellites was not classified. He also noted that LaPaz was not associated with the Army's search, but that "we have the best man available on this project in Dr. Clyde W. Tombaugh, discoverer of the planet Pluto." Colonel Holler's statement was published in many newspapers in late August and September 1954. The *Aviation Week* scoop and Colonel Holler's statement even made it into the *New York Times,* although the story was hidden away on page 35 of the August 29 Sunday edition. The *Times* headlined the story "Earth 'Satellites' Spur Army Study" and appeared to give some credence to the *Aviation Week* report.

The *Aviation Week* story had no basis in fact. LaPaz quickly denied that he had been involved in the discovery of any satellites, although he encouraged the development of a U.S. space program. Tombaugh's search, far from complete in 1954, would continue but turn up no satellites.

Chapter 10
Conclusions and Loose Ends

Despite modifications and adjustments, Clarence Chant's original explanation for the Great Meteor Procession has stood the test of time. On the night of February 9, 1913, witnesses saw meteoric bodies traversing long distances on trajectories nearly parallel to the surface of the Earth. Perhaps no single body traveled the entire 7000-mile distance from Didsbury, Canada, to the South Atlantic Ocean, but, if not, successive meteors took up portions of the journey, amazing witnesses from cold northern prairies to warm tropical seas.

Over the past 110 years, Chant's hypothesis has faced several challenges. Most important was C. C. Wylie's alternative of a detonating fireball accompanied by an ordinary meteor shower. Wylie's explanation did not survive refutations by Alexander Mebane and John O'Keefe.

Nor has Chant's hypothesis been successfully challenged by the alternative of mysterious spaceships. The fireballs advanced in majestic array, but they showed no changes of motion that would imply a piloted craft, no sudden changes of speed or direction. When startled witnesses described individual fireballs within the procession, their descriptions were almost all in accord with bright, if slow, meteors. The reasons some advanced for an artificial nature of the fireballs are simply the large numbers of fireballs, their long track, and their relatively slow apparent motion. The latter two are, however, properties that might be expected of any object in a satellite or near-satellite orbit, natural as well as artificial.

While Alexander Mebane took some UFO reports seriously, even he rejected the spaceship hypothesis for the Great Meteor Procession. He pointed out that the only places from which witnesses actually reported some sort of flying ships were Marshall, Michigan and North Bay, Ontario. In each location, the witnesses were far from the main track of the fireballs. Witnesses closer to the track stated that the objects did indeed look like fireballs. Eyewitness accounts are consistent with a swarm of meteoroids hitting the Earth's atmosphere on a grazing trajectory, resulting in speeds comparable to a satellite's five miles per second.

More than a century after Chant published his explanation for the meteor procession, Chant's hypothesis remains tenable. The original hypothesis may have required a tweak here and a nudge there, but there is no need for outright replacement. The Great Meteor Procession was exceptional, but it was not the handiwork of secret inventors or creatures from beyond our planet.

Are all problems connected with the 1913 meteor procession resolved? No. The papers by O'Keefe and by Beech and Comte presented plausible mechanisms for the production of the meteor procession from the fragmentation of a parent body. However, exactly how that happened in the particular case of the 1913 fireballs has not been fully established. Indeed, it is quite possible that we shall never have the information required to completely understand all aspects of the processes associated with its creation.

Great Meteor Procession

Other loose ends are the meteors which don't fit the Chant procession. In chapter 3, we mentioned daylight observations of objects from the Toronto area on the afternoon after the procession. However, those reports we found to be lacking in details. Chant, however, noted that there were other reports sent to him which involved meteors that seemed real, but which did not obviously belong to the great procession. Mebane added a few additional aberrant observations to the mix in his 1956 paper. What do we make of them?

O'Keefe's (1961) investigation makes a strong case that no stream of meteors circled the Earth at low altitude to appear an hour and a half later over the western United States. Mebane (1956) had, however, argued that there might have been a second procession of meteors following the general path of the first, but five hours later, near 2:30 am (eastern time) on the following morning. The evidence for this second procession is not very strong. Mebane based its existence upon reports by four observers, but three of them were recalling things seen four decades before; only one report collected by Chant was obtained soon after the event. How much weight should we give to observations remembered after four decades? Mebane also had to assume, plausibly but without certainty, that one of the three observers misdated an observation by one day. There just doesn't seem to be enough in the way of reliable observations to establish that the proposed second procession happened, nor does it seem likely that observations adequate to resolve the question can be found now (although I would never discourage anyone from seeking undiscovered observations).

Great Meteor Procession

What about Mebane's (1956) proposal that some witnesses saw meteors going in the same direction as the Great Procession, at the same time as the Procession, but displaced spatially from the main procession? On this point, he is on firmer ground.

Mebane suspected that not all observations of meteors at the time of the Great Procession could be well-fitted with a single line on the map. In support of this, Mebane called attention to observations made in Alpena, in northern Michigan: "The single-file procession across Alpena must be distinguished from the less regular procession (the 2 large fireballs?) that passed to the south of Alpena; and since both of these evidently crossed the lake to Ontario, it must be suspected that Chant erred in fitting all of his observations to a single line." Mebane identified groups of observations from different locations along the Chant track which might refer to distinct fireball groups which traversed only a part of the full route of the Great Meteor Procession. Perhaps glowing fragments were unusually sparse in the vicinity of New York City, explaining why observations from that area were rare.

We conclude with something perhaps more sociological than meteoritic which occurred to Mebane as he completed his researches. He wrote that the most remarkable aspect of his investigation "was its revelation of the extreme spottiness with which celestial phenomena are observed." He noted that one newspaper in the thumb area of Michigan reported that the meteors caused a sensation, while newspapers in eight surrounding towns made no mention at all of the procession. We are fortunate that at least some people paid attention to what they saw that night. That we are able to say anything about

Great Meteor Procession

the Great Meteor Procession depends, after all, not only upon investigators such as Chant, O'Keefe, and Mebane, but upon the witnesses who took the time and made the effort to put their observations on record.

Glossary

Apogee: The point in the orbit of a satellite when it is farthest from the Earth.

Asteroid: A rocky, icy, or metallic object, usually orbiting the sun, but smaller than a planet, is called an asteroid. They can be as large as Ceres, which is about 600 miles (1000 km) across and is sometimes considered a dwarf planet. Some limit the term asteroid to objects a meter or more in diameter. The vast majority of asteroids are a part of our solar system. However, on rare occasions an asteroid is discovered which has come into our solar system from interstellar space.

Bolide: Another term for a fireball, the word bolide is sometimes reserved for those fireballs which explode toward the end of their flight.

Escape speed: This is the speed needed for a projectile to become gravitationally unbound from an object. For the Earth, the escape speed from the surface is 7.0 miles per second or 11.2 kilometers per second. That corresponds to about 25,000 miles per hour. That can be compared with the speed of a satellite in low-Earth orbit. A satellite circling the Earth at an altitude of a few hundred miles will move with a slower speed, close to 4.8 miles per second (7.8 kilometers per second). That corresponds to about 17,500 miles per hour.

Great Meteor Procession

Fireball: As the word indicates, a fireball looks like a bright moving ball of fire or light. Ordinarily, it can be regarded as an unusually bright meteor.

Meteor: Originally used to describe a wide range of atmospheric phenomena, the term meteor is now usually restricted to the luminous display (a shooting star or fireball) created when a meteoroid plunges into the atmosphere at a high speed. In earlier times, the word meteor was sometimes also applied to the solid object producing the light, but recently the term has been used in the more restrictive manner.

Meteorite: If the solid object giving rise to a meteor reaches the ground, the surviving piece or pieces are called meteorites.

Meteoroid: A meteoroid is a small rocky, icy, or metallic object (up to about a meter across) moving through space. When a meteoroid enters the Earth's atmosphere, it creates a meteor. A typical shooting star is caused by a small meteoroid, perhaps the size of a grain of sand or a small pebble.

Perigee: The point in its orbit when a satellite is closest to the Earth.

Radiant: The radiant of a meteor shower is the point in the sky from which meteors belonging to the shower appear to diverge.

Shooting Star: A common term for an ordinary meteor.

Bibliography

Beal, W. J., 1915, History of the Michigan Agricultural College and Biographical Sketches of Trustees and Professors (Agricultural College: East Lansing).

Beech, M. 1989, *The Great Meteor of 18th August 1783*, Journal of the British Astronomical Association **99**, 130.

Beech, M. and Comte, M., 2018, *The Chant Meteor Procession of 1913 – Towards a Descriptive Model*, American Journal of Astronomy and Astrophysics, **6**, 31.

Bertram, Dean, 2006, Flying Saucer Culture: The History of American UFO Belief (PhD Dissertation: University of Sydney).

Blagden, Charles, 1784, *An Account of Some Late Fiery Meteors; With Observations*, Philosophical Transactions of the Royal Society of London, **74**, 201.

Burns, G. J., 1913a, *An Extraordinary Meteoric Display*, Journal of the British Astronomical Association, **24**, 109.

Burns, G. J., 1913b, *Report of the Meeting of the Association*, Journal of the British Astronomical Association, **24**, 147.

Carvallo, Tiberius, 1784, *Description of a Meteor, Observed Aug. 18, 1783*, Philosophical Transactions of the Royal Society of London, **74**, 108.

Chant, Clarence A., 1913a, *An Extraordinary Meteoric Display*, Journal of the Royal Astronomical Society of Canada, **7**, 145.

Chant, Clarence A., 1913b, *Further Information Regarding the Meteoric Display of February 9 1913*, Journal of the Royal Astronomical Society of Canada, 7, 438.

Clark, Jerome, 2018, The UFO Encyclopedia (Detroit: Omnigraphics), Kindle edition, 781.

Coffin, James H., 1869, *The Orbit and Phenomena of a Meteoritic Fire-Ball, seen July 20, 1860*, Smithsonian Contributions to Knowledge, **16**, 1.

Collins, Curt, 2019, *Fantastic Universe: UFOs and Civilian Saucer Intelligence,* The Saucers that Time Forgot, https://thesaucersthattimeforgot.blogspot.com/2019/07

Ceplecha, Z., 1994, *Earth-grazing daylight fireball of August 10, 1972*, Astronomy & Astrophysics, **283**, 287.

Davidson, M. A. 1913, *Report of the Meeting of the Association,* Journal of the British Astronomical Association, **24**, 148.

Denning, W. F., 1913, *Notes on the Great Meteoric Stream of 1913, February 9th*, Journal of the Royal Astronomical Society of Canada, 7, 404.

Denning, W. F., 1915, *The Great Meteoric Stream of February 9, 1913*, Journal of the Royal Astronomical Society of Canada, **9**, 287.

Denning, W. F., 1916, *Great Meteoric Stream of February 9th, 1913*, Journal of the Royal Astronomical Society of Canada, **10**, 294.

Edwards, Frank, 1961, *The Great Sky Procession of 1913*, Fate Magazine, December issue.

Edwards, Frank, 1964, *Express Train in the Sky*, Strange World (New York: Ace Books) 88.

Fedorets, G., et al. 2020, *Establishing Earth's Minimoon Population through Characterization of Asteroid 2020 CD_3*, Astronomical Journal, **160**, 277.

Fisher, Willard J., 1928, *Remarks on the meteoric procession of 1913 February 9*, Popular Astronomy, **36**, 398.

Granvik, M., Vaubaillon, J., and Jedicke, R., 2012, *The Population of Natural Earth Satellites*, Icarus, **218**, 262.

Hartmann, William K., 1968, *Process of Perception, Conception, and Reporting*, in in The Scientific Study of Unidentified Flying Objects, Edward U. Condon, (Colorado Associated University Press and E. P. Dutton & Co.) 567.

Hoffmeister, Cuno, 1937, Die Meteore, (Leipzig: Akademische Verlagsgesellschaft).

Keel, John A., 1970, Operation Trojan Horse (New York: G. P. Putnam's Sons).

Kwiatkowski T. et al., *2009, Photometry of 2006 RH_{120}: an Asteroid Temporary Captured into a Geocentric Orbit*, Astronomy & Astrophysics, **495**, 967.

LaPaz, Lincoln, 1956, *The Canadian Fireball Procession of 1913 February 9*, Meteoritics, **1**, 402.

LaPaz, Lincoln and LaPaz, Jean, 1961, Space Nomads (New York: Holiday House).

Ley, Willy, 1951, *The Meteoric Stream,* Galaxy, September issue, 106.

Longo, M.J. and Morris, R., 1986, *A Sensitive Radar Search for Small Natural Satellites of the Earth,* Astronomical Journal, **91**, 1238.

Lyman, Chester S., 1860, *The Meteor of July 20, 1860,* American Journal of Science and Arts, **30**, 293.

Marvin, U. B., 1993, *The Meteoritical Society: 1933 to 1993,* Meteoritics **28**, 261.

Mebane, Alexander D., 1953, *The "Great Fireball Procession" of 1913* (Letter), Science **118**, 725.

Mebane, Alexander D., 1956, *Observations of the Great Fireball Procession of 1913 February 9, Made in the United States,* Meteoritics, **1**, 405.

Michel, Aimé, 1958, Flying Saucers and the Straight-Line Mystery (New York: S. G. Phillips).

Monck, W. H. S., 1914, *The Great Meteor of 9th February, 1913,* Journal of the Royal Astronomical Society of Canada, **8**, 112.

O'Keefe, John A., 1959, *The Radiant and Orbit of the Meteors of February 9, 1913,* Journal of the Royal Astronomical Society of Canada, **53**, 59.

O'Keefe, John A., 1960, *Tektites and the Cyrillid Shower*, Astronomical Journal, **65**, 495.

O'Keefe, John A., 1961, *Tektites as Natural Earth Satellites*, Science, **133**, 562.

O'Keefe, John A., 1963, *The Origin of Tektites,* Tektites, ed, J. A, O'Keefe (Chicago: University of Chicago Press), 167.

O'Keefe, John A., 1968, *New Data on Cyrillids*, Journal of the Royal Astronomical Society of Canada, **62**, 97.

Olson, Donald, et al. 2010, *Walt Whitman's Year of Meteors,* Sky and Telescope, **120**, 28.

Olson, Donald, 2013, Celestial Sleuth: Using Astronomy to Solve Mysteries in Art, History and Literature (New York: Springer).

Olson, Donald and Hutcheon, Steve, 2013, *The Great Meteor Procession of 1913*, Sky and Telescope, **125**, 32.

Olson, Donald, 2018, Further Adventures of the Celestial Sleuth: Using Astronomy to Solve More Mysteries in Art, History, and Literature (Chichester: Springer Praxis Books).

Olson, Donald, 2022, Investigating Art, History, and Literature with Astronomy: Determining Time, Place, and Other Hidden Details Linked to the Stars (Chichester: Springer Praxis Books).

Pickering, William H., 1922, *The Meteoric Procession of February 9, 1913, Part I,* Popular Astronomy, **30**, 632.

Pickering, William H., 1923a, *The Meteoric Procession of February 9, 1913, Part II*, Popular Astronomy, **31**, 96.

Pickering, William H., 1923b, *The Meteoric Procession of February 9, 1913, Part III*, Popular Astronomy, **31**, 443.

Pickering, William H., 1923c, *The Meteoric Procession of February 9, 1913, Part IV*, Popular Astronomy, **31**, 501.

Rawcliffe R. D. et al., 1974, *Meteor of August 10, 1972*, Nature, **247**, 449.

Shober, P.M. et al., 2019, *Identification of a Minimoon Fireball*, Astronomical Journal, **158**, 183.

Swords, Michael, et al. 2012, UFOs and Government (San Antonio: Anomalist Books).

Tombaugh, C. W., Smith, Bradford A., and Capen, Charles F., Jr., 1957, *Search for small Satellites of the Moon during the Total Lunar Eclipse of November 18, 1956*, Publications of the Astronomical Society of the Pacific, **69**, 400.

Tombaugh, C. W., Robinson, J.C., Smith, B.A., and Murrell, A.S., 1959, The Search for Small Natural Earth Satellites; Final Technical Report (New Mexico State University: Physical Science Laboratory).

Vallée, Jacques, 1965, Anatomy of a Phenomenon (New York: Ace Books).

Wylie, Charles C., 1939, *The Radiant and Orbit of the Meteors of February 9, 1913*, Popular Astronomy, **47**, 291.

Wylie, Charles C., 1940, *Psychological Errors in Meteor Work,* Popular Astronomy, **47**, 206.

Wylie, Charles C., 1953a, *Those Flying Saucers*, Science, **118**, 125.

Wylie, Charles C., 1953b, *The "Great Fireball Procession" of 1913* (Letter), Science, **118**, 726.

Index

Airships, 12, 40, 105-107
Alberta, 63, 69, 85
Army Office of Ordnance Research, 73, 116
Arnold, Kenneth, 105
Asteroid (definition), 122
Australia, 85
Aviation Week, 115, 116
Beal, William, iv
Beech, Martin, 78, 79, 84, 118
Bellusia, 31, 61, 63
Bermuda, 9, 14, 19, 20, 25, 30, 35, 38, 40, 113
Berthold Vinnen, 67
Blagden, Charles, 84
Bolide (definition), 122
Bolton, James, 18
Bond, G. P., 93, 94, 99
Boyd, Lyle, 114, 115
Brazil, 31, 67, 68
Cavallo, Tiberius, 81, 83, 84
Ceres, 122
Chant, Clarence, iv, 2, 5, 6, 8-10, 14, 17, 21, 23-25, 27, 28, 29, 33-36, 38-40, 42, 43, 49, 50, 53-55, 57-61, 63, 65, 67, 69, 94, 100, 107, 113, 117, 119-121
Church, Frederic, 67, 97, 98
Civilian Saucer Intelligence, 111, 112
Coffin, James, 98-104
Comte, Mark, 78, 79, 118

Condon, Edward, 114
Cyril, Alexandrine saint, 65-67
Cyrillids, 65
Davidson, M., 24, 28, 34
Day the Earth Stood Still, The, 106
Denning, W. F., iv, 2, 24, 28-32, 39, 67
Didsbury, 63, 68, 117
Earth-grazing meteor, 25-28, 78, 79, 81, 84, 85, 103, 118
Edwards, Frank, 108, 109
Escape speed (definition), 122
Fate, 106, 108
Fireball (definition), 123
Fireball 1783, 81-86
Fireball 1860, 89-104
Fireball 1972, 84-86, 103
Fireball 2017, 86
Fisher, Willard, 33, 34, 36
Fort, Charles, 2, 37, 38, 40, 41, 56, 105, 106, 112
Geminid meteor shower, 43, 44, 47
Green fireballs, 111, 113, 114
Hahn, Gustav, 9
Haight, Walter, 14, 15
Hartmann, William, 114, 115
Holler, Walker, 116
Hutcheon, Steve, 67
Hynek, J. Allen, 111, 112
J. C. Vinnen, 68, 69

Keel, John, 109
Keyhoe, Donald, 106, 115
LaPaz, Jean, 60
LaPaz, Lincoln, 53, 55, 59-61, 69, 110-116
Leonid meteor shower, 47
Ley, Willy, 107
Longo, M., 75
Lovecraft, H. P., 107
Lowell Observatory, 72, 74
Lyman, Chester, 98, 100
Mebane, Alexander, iv, 2, 41, 51, 54-60, 62-64, 67, 69, 110-121
Menzel, Donald, 113, 114, 115
Meteor (definition), 123
Meteorite (definition), 123
Meteoroid (definition), 123
Michel, Aimé, 112
Michigan State University, iii
Mir space station, 81
Monck, W. H. S., 28
Morris, R., 75
Newlands, 31, 32, 38, 61, 63
North American Aerospace Defense Command, 75
O'Keefe, John, iv, 2, 61-67, 71, 77, 114, 117-119, 121
Olson, Donald, 67, 91, 92, 97
Ormiston, J. T., 19
Palmer, Raymond, 106
Perseid meteor shower, 43, 44, 46, 47, 65
Pickering, William, 32, 33, 36, 40, 67
Porter, A. Y., 31

Pruett, J. H., 51
Radiant (definition), 43, 123
Roswell, NM, 105
Sandby, Paul, 83
Saskatchewan, 9, 14, 25, 31, 34, 38, 40, 49, 50, 113
Satellite, 3, 25, 28, 29, 31, 33, 61-63, 70, 71, 73-76, 81, 86, 108, 115, 116, 118, 122
Shober, P. M., 85, 86
Shooting star, 25, 123
Spaceships, 41, 105-108, 114, 117
Sputnik, 29, 115
Steiger, Brad, 108
Tektites, 65, 66
Thing from Another World, The, 106
Tombaugh, Clyde, 71-75, 115, 116
Toronto, 1, 5, 7-9, 14-16, 18, 19, 21, 24, 30, 31, 38, 39, 40, 49-53, 58, 62, 64, 78, 119
Unidentified Flying Objects, 2, 50, 55, 103, 105, 108-112, 114, 115, 118, 125
Valleé, Jacques, 108
Waddell, W. W., 31
Wells, H. G., 29
White Sands, New Mexico,, 72, 73
Whitman, Walt, 89-92, 97
Winter, W. R., 19, 20

Wylie, C. C., 2, 41-43, 49-53, 56, 58, 60-62, 71, 94, 112-115, 117

Year of Meteors, 89-91, 97, 103

Zafra, 21

www.ingramcontent.com/pod-product-compliance
Lightning Source LLC
Chambersburg PA
CBHW071410210526
45465CB00001B/330